AF522366

RESOURCE GEOGRAPHY

RESOURCE GEOGRAPHY

Dr. Anoop Kumar Singh

RANDOM PUBLICATIONS

NEW DELHI - 110 002 (INDIA)

Resource Geography

ISBN 978-93-51117-41-4

Published in 2015 in India by
RANDOM PUBLICATIONS
4376-A/4B, Gali Murari Lal, Ansari Road
New Delhi-110 002
Phone: +9111-43580356, 23289044
E-mail: randomexports@gmail.com; sales@randompublications.com;
info@randompublications.com

Reprinted 2024

Type Setting by: Friends Media, Delhi-110089
Printed at : Replika Press Pvt. Ltd.

Preface

A resource is a source or supply from which benefit is produced. Typically resources are materials, energy, services, staff, knowledge, or other assets that are transformed to produce benefit and in the process may be consumed or made unavailable. Benefits of resource utilization may include increased wealth, meeting needs or wants, proper functioning of a system, or enhanced well being. From a human perspective a natural resource is anything obtained from the environment to satisfy human needs and wants. From a broader biological or ecological perspective a resource satisfies the needs of a living organism. Everything available in our environment which can be used to satisfy our needs, provided, it is technologically accessible, economically feasible and culturally acceptable can be termed as 'Resource'. We transform things into resources with the help of nature, technology and institutions. The process of transformation of things involves an inter-dependent relationship between these. Human beings interact with nature through technology and create institutions to accelerate their economic development.

The geographically informed person must understand that a "resource" is a cultural concept. A resource is any physical material constituting part of Earth that people need and value. Natural materials become resources when humans value them. The uses and values of resources change from culture to culture and from time to time. Resources are spatially distributed varying in quantity and quality. Some resources are finite, while others can be replenished at varying rates. However, humans need to balance short-term rates of use against long-term availability to ensure a sustainable future. Therefore, Standard 16 contains these themes: Types and Meanings of Resources, Location and Distribution of Resources, and Sustainable Resource Use and Management. Three basic resources—land, water, and air—are essential to survival. The characteristics and quantity of a resource are defined by whether it is a renewable, nonrenewable, or flow resource. Renewable resources can be replenished if their environments remain intact. Nonrenewable resources can be extracted and used only once. Flow resources, such as water, wind, and sunlight, must be used when and where they occur because they are neither renewable nor nonrenewable. Resource location influences the distribution of people and their activities. People settle where they can make a living and where the needed resources are available—resources such as fertile soils, potable water, fuel, and building materials. The patterns of population distribution resulting from the relationship between resources and employment change as needs and technologies change. New technology alters how people appraise resources, influences where they live and work, and affects how economic systems adapt. Three basic resources—land, water, and air—are essential to survival. In countries where there is a rapid population growth and/or consumption of resources, there is a danger of a fall in their standards of living.

The present book is an essential resource for people engaged in the use and management of natural resources, and for those who seek professional training in the field.

I thank all members of my team who have helped in the preparation of the book. My special thanks go to "Random Publications" who have published the book.

— Dr. Anoop Kumar Singh

Contents

Chapter 1

Resource

A resource is a source or supply from which benefit is produced. Typically resources are materials, energy, services, staff, knowledge, or other assets that are transformed to produce benefit and in the process may be consumed or made unavailable. Benefits of resource utilization may include increased wealth, meeting needs or wants, proper functioning of a system, or enhanced well being. From a human perspective a natural resource is anything obtained from the environment to satisfy human needs and wants. From a broader biological or ecological perspective a resource satisfies the needs of a living organism.

The concept of resources has been applied in diverse realms, including with respect to economics, biology and ecology, computer science, management, and human resources, and is linked to the concepts of competition, sustainability, conservation, and stewardship. In application within human society, commercial or non-commercial factors require resource allocation through resource management.

Resources have three main characteristics: utility, limited availability, and potential for depletion or consumption. Resources have been variously categorized as biotic versus abiotic, renewable versus non-renewable, and potential versus actual, along with more elaborate classification.

Economic Resources

In economics a resource is defined as a service or other asset used to produce goods and services that meets human needs and wants. Economics itself has been defined as the study of how society manages its scarce resources. Classical economics recognizes three categories of resources, also referred to as factors of production: land, labour, and capital. Land includes all natural resources and is viewed as both the site of production and the source of raw materials. Labour or human resources consists of human effort provided in the creation of products, paid in wage. Capital consists of human-made goods or means of production (machinery, buildings, and other infrastructure) used in the production of other goods and services, paid in interest.

Biological Resources

In biology and ecology a resource is defined as a substance that is required by a living organism for normal growth, maintenance, and reproduction. Resources, such as food, water, or nesting sites, can be consumed by an organism and, as a result, become unavailable to other organisms.

Economic versus Biological Resources

There are three fundamental differences between economic versus ecological views:

1) the economic resource definition is human-centred (anthropocentric) and the biological or ecological resource definition is nature-centred (biocentric or ecocentric);
2) the economic view includes desire along with necessity, whereas the biological view is about basic biological needs; and
3) economic systems are based on markets of currency exchanged for goods and services, whereas biological systems are based on natural processes of growth, maintenance, and reproduction.

Land or Natural Resources

Natural resources are derived from the environment. Many natural resources are essential for human survival, while others are used for satisfying human desire. Conservation is the management of natural resources with the goal of sustainability. Natural resources may be further classified in different ways.

Resources can be categorized on the basis of origin:

- Abiotic resources comprise non-living things (e.g., land, water, air and minerals such as gold, iron, copper, silver).
- Biotic resources are obtained from the biosphere. Forests and their products, animals, birds and their products, fish and other marine organisms are important examples. Minerals such as coal and petroleum are sometimes included in this category because they were formed from fossilized organic matter, though over long periods of time.

Natural resources are also categorized based on the stage of development:

- Potential Resources are known to exist and may be used in the future. For example, petroleum may exist in many parts of India and Kuwait that have sedimentary rocks, but until the time it is actually drilled out and put into use, it remains a potential resource.
- Actual resources are those that have been surveyed, their quantity and quality determined, and are being used in present times. For example, petroleum and natural gas is actively being obtained from the Mumbai High Fields. The development of an actual resource, such as wood processing depends upon the technology available and the cost involved. That part of the actual resource that can be developed profitably with available technology is called a reserve resource, while that part that can not be developed profitably because of lack of technology is called a stock resource.

Natural resources can be categorized on the basis of renewability:

- Non-renewable Resources are formed over very long geological periods. Minerals and fossils are included in this category. Since their rate of formation is extremely slow, they cannot be replenished, once they are depleted. Out of these, the metallic minerals can be re-used by recycling them, but coal and petroleum cannot be recycled.
- Renewable resources, such as forests and fisheries, can be replenished or reproduced relatively quickly. The highest rate at which a resource can be used sustainably is the sustainable yield. Some resources, like sunlight, air, and wind, are called perpetual resources because they are available continuously, though at a limited rate. Their quantity is not affected by human consumption. Many renewable resources can be depleted by human use, but may also be replenished, thus maintaining a flow. Some of these, like agricultural crops, take a short time for renewal; others, like water, take a comparatively longer time, while still others, like forests, take even longer.

Dependent upon the speed and quantity of consumption, overconsumption can lead to depletion or total and everlasting destruction of a resource. Important examples are agricultural areas, fish and other animals, forests, healthy water and soil, cultivated and natural landscapes. Such conditionally renewable resources are sometimes classified as a third kind of resource, or as a subtype of renewable resources.

Conditionally renewable resources are presently subject to excess human consumption and the only sustainable long term use of such resources is within the so-called zero ecological footprint, wherein human use less than the Earth's ecological capacity to regenerate.

Natural resources are also categorized based on distribution:

- Ubiquitous Resources are found everywhere (e.g., air, light, water).
- Localized Resources are found only in certain parts of the world (e.g., copper and iron ore, geothermal power).

On the basis of ownership, resources can be classified as individual, community, national, and international.

Labour or Human Resources

Human beings, through the labour they provide and the organizations they staff, are also considered to be resources. The term human resources can also be defined as the skills, energy, talent, abilities, and knowledge that are used for the production of goods or the rendering of services.

In a project management context, human resources are those employees responsible for undertaking the activities defined in the project plan.

Capital or Infrastructure

In economics, capital refers to already-produced durable goods used in production of goods or services. As resources, capital goods may or may not be significantly consumed, though they may depreciate in the production process and they are typically of limited capacity or unavailable for use by others.

Tangible versus Intangible Resources

Whereas, tangible resources such as equipment have actual physical existence, intangible resources such as corporate images, brands and patents, and other intellectual property exist in abstraction.

Generally the economic value of a resource is controlled by supply and demand. Some view this as a narrow perspective on resources because there are many intangibles that cannot be measured in money. Natural resources such as forests and mountains have aesthetic value. Resources also have an ethical value.

Resource Use and Sustainable Development

Typically resources cannot be consumed in their original form, but rather through resource development they must be processed into more usable commodities. With increasing population, the demand for resources is increasing. There are marked differences in resource distribution and associated economic inequality between regions or countries, with developed countries using more natural resources than developing countries. Sustainable development is a pattern of resource use, that aims to meet human needs while preserving the environment.

Natural Resource

Natural resources occur naturally within environments that exist relatively undisturbed by humanity, in a natural form. A natural resource is often characterized by amounts of biodiversity and geodiversity existent in various ecosystems.

Natural resources are derived from the environment. Some of them are essential for our survival while most are used for satisfying our needs. Natural resources may be further classified in different ways.

Natural resources are materials and components (something that can be used) that can be found within the environment. Every man-made product is composed of natural resources (at its fundamental level). A natural resource may exist as a separate entity such as fresh water, and air, as well as a living organism such as a fish, or it may exist in an alternate form which must be processed to obtain the resource such as metal ores, oil, and most forms of energy.

There is much debate worldwide over natural resource allocations, this is partly due to increasing scarcity (depletion of resources) but also because the exportation of natural resources is the basis for many economies (particularly for developed nations).

Some natural resources such as sunlight and air can be found everywhere, and are known as ubiquitous resources. However, most resources only occur in small sporadic areas, and are referred to as localized resources. There are very few resources that are considered inexhaustible (will not run out in foreseeable future) – these are solar radiation, geothermal energy, and air (though access to clean air may not be). The vast majority of resources are exhaustible, which means they have a finite quantity, and can be depleted if managed improperly.

Classification

There are various methods of categorizing natural resources, these include source of origin, stage of development, and by their renewability.

These classifications are described below. On the basis of origin, resources may be divided into:

- *Biotic* – Biotic resources are obtained from the biosphere (living and organic material), such as forests and animals, and the materials that can be obtained from them. Fossil fuels such as coal and petroleum are also included in this category because they are formed from decayed organic matter.
- *Abiotic* – Abiotic resources are those that come from non-living, non-organic material. Examples of abiotic resources include land, fresh water, air and heavy metals including ores such as gold, iron, copper, silver, etc.

Considering their stage of development, natural resources may be referred to in the following ways:

- *Potential resources* – Potential resources are those that exist in a region and may be used in the future. For example petroleum occurs with sedimentary rocks in various regions, but until the time it is actually drilled out and put into use, it remains a potential resource.
- *Actual resources* – Actual resources are those that have been surveyed, their quantity and quality determined and are being used in present times. The development of an actual resource, such as wood processing depends upon the technology available and the cost involved.
- *Reserve resources* – The part of an actual resource which can be developed profitably in the future is called a reserve resource.
- *Stock resources* – Stock resources are those that have been surveyed but cannot be used by organisms due to lack of technology. For example: hydrogen.

Renewable Resource

A renewable resource is an organic natural resource that can replenish in due time compared to the usage, either through biological reproduction or other naturally recurring processes. Renewable resources are a part of Earth's natural environment and the largest components of its ecosphere. A positive life cycle assessment is a key indicator of a resource's sustainability.

Definitions of renewable resources may also include agriculture production, as in sustainable agriculture and to an extent water resources. In 1962 Paul Alfred Weiss defined Renewable Resources as: *"The total range of living organisms providing man with food, fibres, drugs, etc...*" The term does not refer however to metals, minerals and fossil fuels. While energy per se is not consumed, some include as well the provision of renewable energy by solar, geothermal and wind power stations under the definition.

Food and Water

Water Resources: Water can be considered a renewable material when carefully controlled usage, treatment, and release are followed. If not, it would become a non-renewable resource at that location. For example, groundwater is usually removed from an aquifer at a rate much greater than its very slow natural recharge, and so groundwater is considered non-renewable. Removal of water from the pore spaces may cause permanent compaction (subsidence) that cannot be renewed. 97.5% of the water on the Earth is salt water, and 3% is fresh water; slightly over two thirds of this is frozen in glaciers and polar ice caps. The remaining unfrozen freshwater is found mainly as groundwater, with only a small fraction (0,008%) present above ground or in the air.

Water pollution is one of the main concerns regarding water resources. It is estimated that 22% of worldwide water is used in industry. Major industrial users include hydroelectric dams, thermoelectric power plants, which use water for cooling, ore and oil refineries, which use water in chemical processes, and manufacturing plants, which use water as a solvent.

Non Agricultural Food

Food is any substance consumed to provide nutritional support for the body. Most food has its origin in renewable resources. Food is obtained directly from plants and animals.

Wild berries and other fruits, mushrooms, plants, seeds and naturally growing edible resources, still represent a valuable source of nutrition in many countries, especially in rural areas, in fact many wild animals are dependent on wild plants and fruits as a source of food.

Hunting may not be the first source of meat in the modernised world, but it is still an important and essential source for many rural and remote groups. It is also the sole source of feeding for wild carnivores.

Sustainable Agriculture

The phrase sustainable agriculture was coined by Australian agricultural scientist Gordon McClymont. It has been defined as "an integrated system of plant and animal production practices having a site-specific application that will last over the long term". Expansion of agricultural land has an impact on biodiversity and contributes to deforestation. The Food and Agriculture Organisation of the United Nations estimates that in coming decades, cropland will continue to be lost to industrial and urban development, along with reclamation of wetlands, and conversion of forest to cultivation, resulting in the loss of biodiversity and increased soil erosion.Although air and sunlight are available everywhere on Earth, crops also depend on soil nutrients and the availability of water. Monoculture is a method of growing only one crop at a time in a given field, which can damage land and cause it to become either unusable or suffer from reduced yields. Monoculture can also cause the build-up of pathogens and pests that target one specific species. The Great Irish Famine (1845–1849) is a well-known example of the dangers of monoculture.

Crop rotation and long-term crop rotations confer the replenishment of nitrogen through the use of green manure in sequence with cereals and other crops, and can improve soil structure and fertility by alternating deep-rooted and shallow-rooted plants. Other methods to combat lost soil nutrients are returning to natural cycles that annually flood cultivated lands (returning lost nutrients indefinitely) such as the Flooding of the Nile, the long-term use of biochar, and use of crop and livestock landraces that are adapted to less than ideal conditions such as pests, drought, or lack of nutrients.

Agricultural practices are the single greatest contributor to the global increase in soil erosion rates. It is estimated that "more than a thousand million tonnes of southern Africa's soil are eroded every year. Experts predict that crop yields will be halved within thirty to fifty years if erosion continues at present rates." The Dust Bowl phenomenon in the 1930s was caused by severe drought combined with farming methods that did not include crop rotation, fallow fields, cover crops, soil terracing and wind-breaking trees to prevent wind erosion.

The tillage of agricultural lands is one of the primary contributing factors to erosion, due to mechanized agricultural equipment that allows for deep plowing, which severely increases the amount of soil that is available for transport by water erosion. The phenomenon called *Peak Soil* describes how large-scale factory farming techniques are jeopardizing humanity's ability to grow food in the present and in the future. Without efforts to improve soil management practices, the availability of arable soil will become increasingly problematic.

Methods to combat erosion include no-till farming, using a keyline design, growing wind breaks to hold the soil, and widespread use of compost. Chemical fertilizer and pesticides can also have an effect of soil erosion, which can contribute to soil salinity and prevent other species from growing. Phosphate is a primary component in the chemical fertilizer applied most commonly in modern agricultural production. However, scientists estimate that rock phosphate reserves will be depleted in 50–100 years and that *Peak Phosphate* will occur in about 2030.

Industrial processing and logistics also have an effect on agriculture's sustainability. The way and locations crops are sold requires energy for transportation, as well as the energy cost for materials, labour, and transport. Food sold at a local location, such a farmers' market, have reduced energy overheads.

Overview Non-Food Resources

The most important renewable resource at large has been and is still wood provided by means of forestry, which is used for construction, housing and firewood since ancient times. Plants provide the main sources for renewable resources, the main distinction is made between energy crops and Non-food crops. A large variety of lubricants, industrially used vegetable oils, textiles and fibre made e.g. of cotton, copra or hemp, paper derived from wood, rags or grasses, bioplastic are based on plant renewable resources. A large variety of chemical base products like latex, ethanol, resin, sugar and starch can be provided with plant renewables. Animal based renewables include fur, leather, technical fat and lubricants and further derived products, as e.g. animal glue, tendons, casings or in historical times ambra and baleen provided by whaling.

With regard to pharmacy ingredients and legal and illegal drugs, plants are important sources, however e.g. venom of snakes, frogs and insects has been a valuable renewable source of pharmacological ingredients. Before GMO production set in, insulin and important hormones ware based on animal sources. Feathers an important byproduct of poultry farming for food is still being used as filler and as base for keratin in general. Same applies for the chitin produced in farming Crustaceans which may be used as base of chitosan. The most important part of the human body used for non medical purposes is human hair as for artificial hair integrations, which is being traded world wide.

Historical Role

Historically, renewable resources like firewood, latex, guano, charcoal, wood ash, plant colours as indigo, and whale products have been crucial for human needs but failed to supply demand in the begin of the industrial era. Early modern times faced large problems with overuse of renewable resources as in deforestation, overgrazing or overfishing.

Besides fresh meat and milk, which is as a food item not topic of this section, livestock farmers and artisans used further animal ingredients as tendons, horn, bones, bladders. Complex technical constructions as the composite bow were based on combination of animal and plant based materials. The current distribution conflict between biofuel and food production is being described as Food vs. fuel. Conflicts between food needs and usage, as supposed by fief obligations were in so far common in historical times as well. However, a significant percentage of (middle European) farmers yields went into livestock, which provides as well organic fertilizer. Oxen and horses were important for transportation purposes, drove engines as e.g. in treadmills.

Other regions solved the transportation problem with terracing, urban and garden agriculture. Further conflicts as between forestry and herding, or (sheep) herders and cattle farmers led to various solutions. Some confined wool production and sheep to large state and nobility domains or outsourced to professional shepherds with larger wandering herds.

The British Agricultural Revolution was mainly based on a new system of crop rotation, the four-field rotation. British agriculturist Charles Townshend recognized the invention in Dutch Waasland and popularized it in the 18th century UK, George Washington Carver in the USA. The system used wheat, turnips amd barley and introduced as well clover. Clover is able to fix nitrogen from air, a practically non exhaustive renewable resource, into fertilizing compounds to the soil and allowed to increase yields by large. Farmers opened up a fodder crop and grazing crop. Thus livestock could to be bred year-round and winter culling was avoided. The amount of manure rose and allowed more crops but to refrain from wood pasture.

Early modern times and the 19th century saw the previous resource base partially replaced respectively supplemented by large scale chemical synthesis and by the use of fossil and mineral resources respectively. Besides the still central role of wood, there is a sort of renaissance of renewable products based on modern agriculture, genetic research and extraction technology. Besides fears about an upcoming global shortage of fossil fuels, local shortages due to boycotts, war and blockades or just transportation problems in remote regions have contributed to different methods of replacing or substituting fossil resources based on renewables.

Challenges

The use of certain basically renewable products as in TCM endangers various species. Just the black market in rhinoceros horn reduced the world's rhino population by more than 90 percent over the past 40 years.

Renewables Used for Autarky Approaches

The success of the German chemical industry till World War I was based on the replacement of colonial products. The predecessors of IG Farben dominated the world market for synthetic dyes at the beginning of the 20th century and had an important role in artificial pharmaceuticals, photographic film, agricultural chemicals and electrochemicals.

However the former Plant breeding research institutes took a different approach. After the loss of the German colonial empire, important players in the field as Erwin Baur and Konrad Meyer switched to using local crops as base for economic autarky. Meyer as a key agricultural scientist and spatial planner of the Nazi era managed and lead Deutsche Forschungsgemeinschaft resources and focused about a third of the complete research grants in Nazi Germany on agricultural and genetic research and especially on resources needed in case of a further German war effort. A wide array of agrarian research institutes still existing today and having importance in the field was founded or enlarged in the time.

There were some major failures as trying to e.g. grow frost resistant olive species, but some success in the case of hemp, Linum, rapeseed, which are still of current importance. During the war, German scientists tried to systematically exploit foreign research results in occupied countries. Heinrich Himmler personally supported a research project using Russian Taraxacum (dandelion) species to manufacture natural rubber. The project was conducted using 150 female KZ prisoners and captured Russian scientists kept together as 'Kommando Pflanzenzucht' (Plant breeding command) in a subcamp (SS) of Konzentrationslager Auschwitz led by SS agrarian research officer Joachim Caesar. Rubber dandelions are still of interest, as scientists in the Fraunhofer Institute for Molecular Biology and Applied Ecology (IME) announced 2013 to have developed a cultivar that is suitable for commercial production of natural rubber.

Legal Situation and Subsidies

Several legal and economic means have been used to enhance the market share of renewables. The UK uses Non-Fossil Fuel Obligations

(NFFO), a collection of orders requiring the electricity Distribution Network Operators in England and Wales to purchase electricity from the nuclear power and renewable energy sectors. Similar mechanisms operate in Scotland (the Scottish Renewable Orders under the Scottish Renewables Obligation) and Northern Ireland (the Northern Ireland Non-Fossil Fuel Obligation). In the USA, Renewable Energy Certificates (RECs), use a similar approach. German Energiewende is using fed-in tariffs. An unexpected outcome of the subsidies was the quick increase of pellet by firing in conventional fossil fuel plants (compare Tilbury power stations) and cement works, making wood respectively biomass accounting for about half of Europe's renewable-energy consumption.

Examples of Industrial Use

Bioplastics

Bioplastics are a form of plastics derived from renewable biomass sources, such as vegetable fats and oils, lignin, corn starch, pea starch or microbiota. The most common form of bioplastic is thermoplastic starch. Other forms include Cellulose bioplastics, biopolyester, Polylactic acid, and bio-derived polyethylene.

The production and use of bioplastics is generally regarded as a more sustainable activity when compared to plastic production from petroleum (petroplastic); however, manufacturing of bioplastic materials is often still reliant upon petroleum as an energy and materials source. Because of the fragmentation in the market and ambiguous definitions it is difficult to describe the total market size for bioplastics, but the global production capacity is estimated at 327,000 tonnes. In contrast, global consumption of all flexible packaging is estimated at around 12.3 million tonnes.

Bioasphalt

Bioasphalt is an asphalt alternative made from non-petroleum based renewable resources. Manufacturing sources of bioasphalt include sugar, molasses and rice, corn and potato starches, and vegetable oil based waste. Asphalt made with vegetable oil based binders was patented by Colas SA in France in 2004.

Renewable Energy

Renewable energy refers to the provision of energy via renewable resources which are naturally replenished fast enough as being used. It includes e.g. sunlight, wind, biomass, rain, tides, waves and geothermal heat. Renewable energy may replace or enhance fossil energy supply various distinct areas: electricity generation, hot water/ space heating, motor fuels, and rural (off-grid) energy services.

Biomass

Biomass is referring to biological material from living, or recently living organisms, most often referring to plants or plant-derived materials.

Sustainable harvesting and use of renewable resources (i.e., maintaining a positive renewal rate) can reduce air pollution, soil contamination, habitat destruction and land degradation. Biomass energy is derived from six distinct energy sources: garbage, wood, plants, waste, landfill gases, and alcohol fuels. Historically, humans have harnessed biomass-derived energy since the advent of burning wood to make fire, and wood remains the largest biomass energy source today.

However, low tech use of biomass, which still amounts for more than 10% of world energy needs may induce indoor air pollution in developing nations and results in between 1.5 million and 2 million deaths in 2000.

The biomass used for electricity generation varies by region. Forest by-products, such as wood residues, are common in the United States. Agricultural waste is common in Mauritius (sugar cane residue) and Southeast Asia (rice husks). Animal husbandry residues, such as poultry litter, are common in the UK. The biomass power generating industry in the United States, which consists of approximately 11,000 MW of summer operating capacity actively supplying power to the grid, produces about 1.4 percent of the U.S. electricity supply.

Biofuel

A biofuel is a type of fuel whose energy is derived from biological carbon fixation. Biofuels include fuels derived from biomass conversion, as well as solid biomass, liquid fuels and various biogases.

Bioethanol is an alcohol made by fermentation, mostly from carbohydrates produced in sugar or starch crops such as corn, sugarcane or switchgrass.

Biodiesel is made from vegetable oils and animal fats. Biodiesel is produced from oils or fats using transesterification and is the most common biofuel in Europe.

Biogas is methane produced by the process of anaerobic digestion of organic material by an aerobes., etc. is also a renewable source of energy.

Biogas

Biogas typically refers to a mixture of gases produced by the breakdown of organic matter in the absence of oxygen. Biogas is produced by anaerobic digestion with anaerobic bacteria or fermentation of biodegradable materials such as manure, sewage, municipal waste, green waste, plant material, and crops. It is primarily methane (CH_4) and carbon dioxide (CO_2) and may have small amounts of hydrogen sulphide (H_2S), moisture and siloxanes.

Natural Fibre

Natural fibres are a class of hair-like materials that are continuous filaments or are in discrete elongated pieces, similar to pieces of thread. They can be used as a component of composite materials. They can also be matted into sheets to make products such as paper or felt. Fibres are of two types: natural fibre which consists of animal and plant fibres, and man made fibre which consists of synthetic fibres and regenerated fibres.

Threats to Renewable Resources

Renewable resources are endangered by non-regulated industrial developments and growth. They must be carefully managed to avoid exceeding the natural world's capacity to replenish them. A life cycle assessment provides a systematic means of evaluating renewability. This is a matter of sustainability in the natural environment.

Overfishing

National Geographic has described ocean over fishing as "simply the taking of wildlife from the sea at rates too high for fished species to replace themselves."

Tuna meat is driving overfishing as to endanger some species like the bluefin tuna. The European Community and other organisations are trying to regulate fishery as to protect species and to prevent their extinctions. The United Nations Convention on the Law of the Sea treaty deals with aspects of overfishing in articles 61, 62, and 65. Examples of overfishing exist in areas such as the North Sea of Europe, the Grand Banks of North America and the East China Sea of Asia.

The decline of penguin population is caused in part by overfishing, caused by human competition over the same renewable resources.

Deforestation

As well as being a renewable resource for fuel and building material, trees protect the environment by absorbing carbon dioxide and by creating oxygen. The destruction of rain forests is one of the critical causes of climate change. Deforestation causes carbon dioxide to linger in the atmosphere. As carbon dioxide accrues, it produces a layer in the atmosphere that traps radiation from the sun. The radiation converts to heat which causes global warming, which is better known as the greenhouse effect.

Deforestation also affects the water cycle. It reduces the content of water in the soil and groundwater as well as atmospheric moisture. Deforestation reduces soil cohesion, so that erosion, flooding and landslides ensue.Rain forest shelter many species and organisms providing local populations with food and other commodities. In this way biofuels may well be unsustainable if their production contributes to deforestation.

Endangered Species

Some renewable resources, species and organisms are facing a very high risk of extinction caused by growing human population and over-consumption. It has been estimated that over 40% of all living species on Earth are at risk of going extinct. Many nations have laws to protect hunted species and to restrict the practice of hunting. Other conservation methods includes restricting land development or creating preserves. The IUCN Red List of Threatened Species is the best-known worldwide conservation status listing and ranking system. Internationally, 199 countries have signed an accord agreeing to create Biodiversity Action Plans to protect endangered and other threatened species.

Non-renewable Resource

A non-renewable resource (also called a finite resource) is a resource that does not renew itself at a sufficient rate for sustainable economic extraction in meaningful human time-frames. An example is carbon-based, organically-derived fuel. The original organic material, with the aid of heat and pressure, becomes a fuel such as oil or gas. Fossil fuels (such as coal, petroleum, and natural gas), and certain aquifers are all non-renewable resources.

In contrast, resources such as timber (when harvested sustainably) and wind (used to power energy conversion systems) are considered renewable resources, largely because their localized replenishment can occur within timeframes meaningful to humans.

Earth Minerals and Ores

Minerals and metal ores are other examples of non-renewable resources. The metals themselves are present in vast amounts in the earth's crust, and their extraction by humans only occurs where they are concentrated by natural geological processes (such as heat, pressure, organic activity, weathering and other processes) enough to become economically viable to extract. These processes generally take from tens of thousands to millions of years, through plate tectonics, tectonic subsidence and crustal recycling.

The localized deposits of metal ores near the surface which can be extracted economically by humans are non-renewable in human time-frames, but on a world scale the most common metal ores as a whole are virtually inexhaustible, in the terms that the amount vastly exceeds human demand or the capacity to mine them, on all time-frames.

There are certain rare earth minerals and elements that are more scarce and exhaustible than others, these are in high demand in manufacturing, particularly for the electronics industry.

Although metal ores are technically non-renewable, similar to rocks and sand, humans are unlikely at the current rate to deplete the world's supply. In this respect, metal ores are considered vastly greater in supply to fossil fuels, because metal ores are formed by crustal-scale processes which make up a much larger portion of the earth's near-surface environment, than those that form fossil fuels. Which are limited to areas where carbon-based life forms flourish, die, and are quickly buried, moreover these fossil fuel forming environments and deposits that we use today occurred extensively in the Carboniferous Period of history.

Fossil Fuel

Natural resources such as coal, petroleum (crude oil) and natural gas take thousands of years to form naturally and cannot be replaced as fast as they are being consumed. Eventually it is considered that fossil-based resources will become too costly to harvest and humanity will need to shift its reliance to other sources of energy. These resources are yet to be named.

An alternative hypothesis is that carbon based fuel is virtually inexhaustible in human terms, if one includes all sources of carbon-based energy such as methane hydrates on the sea floor, which are vastly greater than all other carbon based fossil fuel resources combined. These sources of carbon are also considered non-renewable, although their rate of formation/replenishment on the sea floor is not known. However their extraction at economically viable costs and rates has yet to be determined. At present, the main energy source used by humans is non-renewable fossil fuels. Since the dawn of internal combustion engine technologies in the 17th century, petroleum and other fossil fuels have remained in continual demand. As a result, conventional infrastructure and transport systems, which are fitted to combustion engines, remain prominent throughout the globe. The continual use of fossil fuels at the current rate is believed to increase global warming and cause more severe climate change.

Nuclear Fuel

In 1987, the World Commission on Environment and Development (WCED) an organization set up by but independent from the United Nations classified fission reactors that produce more fissile nuclear fuel than they consume -i.e breeder reactors, and when it is developed, fusion power, among conventional renewable energy sources, such as solar and falling water. The American Petroleum Institute likewise does not consider conventional nuclear fission as renewable, but that breeder reactor nuclear power fuel is considered renewable and sustainable, before explaining that radioactive waste from used spent fuel rods remains dangerous, and so has to be very carefully stored for up to a thousand years. With the careful monitoring of radioactive waste products also being required upon the use of other renewable energy sources, such as geothermal energy.

The use of nuclear technology relying on fission requires Naturally occurring radioactive material as fuel. Uranium, the most common fission fuel, and is present in the ground at relatively low concentrations and mined in 19 countries. This mined uranium is used to fuel energy-generating nuclear reactors with fissionable uranium-235 which generates heat that is ultimately used to power turbines to generate electricity.

Nuclear power provides about 6% of the world's energy and 13–14% of the world's electricity. The expense of the nuclear industry remains predominantly reliant on subsidies and indirect insurance subsidies to continue. Nuclear energy production is associated with potentially dangerous radioactive contamination as it relies upon unstable elements. In particular, nuclear power facilities produce about 200,000 metric tons of low and intermediate level waste (LILW) and 10,000 metric tons of high level waste (HLW) (including spent fuel designated as waste) each year worldwide.

The use of nuclear fuel and the high-level radioactive waste the nuclear industry generates is highly hazardous to people and wildlife. Radiocontaminants in the environment can enter the food chain and become bioaccumulated. Internal or external exposure can cause mutagenic DNA breakage producing teratogenic generational birth defects, cancers and other damage. The United Nations (UNSCEAR) estimated in 2008 that average annual human radiation exposure includes 0.01 mSv (milli-Sievert) from the legacy of past atmospheric nuclear testing plus the Chernobyl disaster and the nuclear fuel cycle, along with 2.0 mSv from natural radioisotopes and 0.4 mSv from cosmic rays; all exposures vary by location. natural uranium radioisotopes in nuclear waste and naturally in the ground emits radiation for the prolonged period of 4.5 billion years or more, and storage has risks of containment. The storage of waste, health implications and dangers of radioactive fuel continue to be a topic of debate, resulting in a controversial and unresolved industry.

Renewable Resources

Natural resources, known as renewable resources, are replaced by natural processes and forces persistent in the natural environment. There are intermittent and reoccurring renewables, and recyclable materials, which are utilized during a cycle across a certain amount of time, and can be harnessed for any number of cycles. The production of goods and services by manufacturing products in economic systems creates many types of waste during production and after the consumer has made use of it. The material is then either incinerated, buried in a landfill or recycled for reuse. Recycling turns materials of value that would otherwise become waste into valuable resources again.

The natural environment, with soil, water, forests, plants and animals are all renewable resources, as long as they are adequately monitored, protected and conserved. Sustainable agriculture is the cultivation of plant and animal materials in a manner that preserves plant and animal ecosystems over the long term. The overfishing of the oceans is one example of where an industry practice or method can threaten an ecosystem, endanger species and possibly even determine whether or not a fishery is sustainable for use by humans. An unregulated industry practice or method can lead to a complete resource depletion.

The renewable energy from the sun, wind, wave, biomass and geothermal energies are based on renewable resources. Renewable resources such as the movement of water (hydropower, tidal power and wave power), wind and radiant energy from geothermal heat (used for geothermal power) and solar energy (used for solar power) are practically infinite and cannot be depleted, unlike their non-renewable counterparts, which are likely to run out if not used sparingly.

The potential wave energy on coastlines can provide 1/5 of world demand. Hydroelectric power can supply 1/3 of our total energy global needs. Geothermal energy can provide 1.5 more times the energy we need. There is enough wind to power the planet 30 times over, wind power could power all of humanity's needs alone. Solar currently supplies only 0.1% of our world energy needs, but there is enough out there to power humanity's needs 4,000 times over, the entire global projected energy demand by 2050.

Renewable energy and energy efficiency are no longer niche sectors that are promoted only by governments and environmentalists. The increasing levels of investment and that more of the capital is from conventional financial actors, both suggest that sustainable energy has become mainstream and the future of energy production, as non-renewable resources decline. This is reinforced by climate change concerns, nuclear dangers and accumulating radioactive waste, high oil prices, peak oil and increasing government support for renewable energy. These factors are commercializing renewable energy, enlarging the market and growing demand, the adoption of new products to replace obsolete technology and the conversion of existing infrastructure to a renewable standard.

Economic Models

In economics, a non-renewable resource is defined as goods, where greater consumption today implies less consumption tomorrow. David Ricardo in his early works analysed the pricing of exhaustible resources, where he argued that the price of a mineral resource should increase over time. He argued that the spot price is always determined by the mine with the highest cost of extraction, and mine owners with lower extraction costs benefit from a differential rent. The first model is defined by Hotelling's rule, which is a 1931 economic model of non-renewable resource management by Harold Hotelling. It shows that efficient exploitation of a nonrenewable and nonaugmentable resource would, under otherwise stable conditions, lead to a depletion of the resource. The rule states that this would lead to a net price or "Hotelling rent" for it that rose annually at a rate equal to the rate of interest, reflecting the increasing scarcity of the resources. The Hartwick's rule provides an important result about the sustainability of welfare in an economy that uses non-renewable source.

However, nearly all metal prices have been declining over time in inflation adjusted terms, because of a number of false assumptions in the above. Firstly, metal resources are non-renewable, but on a world scale, largely inexhaustible. This is because they are present throughout the earth's crust on a vast scale, far exceeding human demand on all time scales. Metal ores however, are only extracted in those areas where nature has concentrated the metal in the crust to a level whereby it is locally economic to extract. This also depends on the available technology for both finding the metal ores as well as extracting them, which is constantly changing. If the technology or demand changes, vast amounts of metal previously ignored can become economically extractable. This is why Ricardo's simplistic notion that the price of a mineral resource should increase over time has in fact turned out to be the opposite, nearly all metal ores have decreased in inflation adjusted prices since well before the early 20th century.

The main reason he was wrong is that he assumed that metals are exhaustible on a world scale, and he also misunderstood the effect of globally competing markets; in human terms the amount of metal in the earth's crust is essentially limitless. It is only in localized areas that metal ores can become depleted, as these local areas compete with extraction costs of resources elsewhere, which does have ramifications for the sustainability of local economies.

Extraction

Resource extraction involves any activity that withdraws resources from nature. This can range in scale from the traditional use of preindustrial societies, to global industry. Extractive industries are, along with agriculture, the basis of the primary sector of the economy. Extraction produces raw material which is then processed to add value. Examples of extractive industries are hunting, trapping, mining, oil and gas drilling, and forestry. Natural resources can add substantial amounts to a country's wealth, however a sudden inflow of money caused by a resource boom can create social problems including inflation harming other industries ("Dutch disease") and corruption, leading to inequality and underdevelopment, this is known as the "resource curse".

Extractive industries represent a large growing activity in many less-developed countries but the wealth generated does not always lead to sustainable and inclusive growth. Extractive industry businesses often are assumed to be interested only in maximizing their short-term value, implying that less-developed countries are vulnerable to powerful corporations. Alternatively, host governments are often assumed to be only maximizing immediate revenue. Researchers argue there are areas of common interest where development goals and business cross. These present opportunities for international governmental agencies to engage with the private sector and host governments through revenue management and expenditure accountability, infrastructure development, employment creation, skills and enterprise development and impacts on children, especially girls and women.

Depletion

In recent years, the depletion of natural resources has become a major focus of governments and organizations such as the United Nations (UN). This is evident in the UN's Agenda 21 Section Two, which outlines the necessary steps to be taken by countries to sustain their natural resources. The depletion of natural resources is considered to be a sustainable development issue. The term sustainable development has many interpretations, most notably the Brundtland Commission's 'to ensure that it meets the needs of the present without compromising the ability of future generations to meet their own needs', however in broad terms it is balancing the needs of the planet's people and species now and in the future. In regards to natural resources, depletion is of concern for sustainable development as it has the ability to degrade current environments and potential to impact the needs of future generations.

Depletion of natural resources is associated with social inequity. Considering most biodiversity are located in developing countries, depletion of this resource could result in losses of ecosystem services for these countries. Some view this depletion as a major source of social unrest and conflicts in developing nations.

At present, with it being the year of the forest, there is particular concern for rainforest regions which hold most of the Earth's biodiversity. According to Nelson deforestation and degradation affect 8.5% of the world's forests with 30% of the Earth's surface already cropped. If we consider that 80% of people rely on medicines obtained from plants and ¾ of the world's prescription medicines have ingredients taken from plants, loss of the world's rainforests could result in a loss of finding more potential life saving medicines.

The depletion of natural resources is caused by 'direct drivers of change' such as Mining, petroleum extraction, fishing and forestry as well as 'indirect drivers of change' such as demography, economy, society, politics and technology. The current practice of Agriculture is another factor causing depletion of natural resources. For example the depletion of nutrients in the soil due to excessive use of nitrogen and desertification The depletion of natural resources is a continuing concern for society. This is seen in the cited quote given by Theodore

Roosevelt, a well-known conservationist and former United States president, was opposed to unregulated natural resource extraction.

Protection

In 1982 the UN developed the World Charter for Nature, which recognised the need to protect nature from further depletion due to human activity. It states that measures need to be taken at all societal levels, from international to individual, to protect nature. It outlines the need for sustainable use of natural resources and suggests that the protection of resources should be incorporated into national and international systems of law. To look at the importance of protecting natural resources further, the World Ethic of Sustainability, developed by the IUCN, WWF and the UNEP in 1990, set out eight values for sustainability, including the need to protect natural resources from depletion. Since the development of these documents, many measures have been taken to protect natural resources including establishment of the scientific field and practice of conservation biology and habitat conservation, respectively.

Conservation biology is the scientific study of the nature and status of Earth's biodiversity with the aim of protecting species, their habitats, and ecosystems from excessive rates of extinction. It is an interdisciplinary subject drawing on science, economics and the practice of natural resource management. The term *conservation biology* was introduced as the title of a conference held at the University of California, San Diego, in La Jolla, California, in 1978, organized by biologists Bruce A. Wilcox and Michael E. Soulé.

Habitat conservation is a land management practice that seeks to conserve, protect and restore, habitat areas for wild plants and animals, especially conservation reliant species, and prevent their extinction, fragmentation or reduction in range.

Management

Natural resource management is a discipline in the management of natural resources such as land, water, soil, plants and animals, with a particular focus on how management affects the quality of life for both present and future generations.

Management of natural resources involves identifying who has the right to use the resources and who does not for defining the boundaries of the resource. The resources are managed by the users according to the rules governing of when and how the resource is used depending on local condition.

A successful management of natural resources should engage the community because of the nature of the shared resources the individuals who are affected by the rules can participate in setting or changing them. The users have the rights to devise their own management institutions and plans under the recognition by the government. The right to resources includes land, water, fisheries and pastoral rights. The users or parties accountable to the users have to actively monitor and ensure the utilisation of the resource compliance with the rules and to impose penalty on those peoples who violates the rules. These conflicts are resolved in a quick and low cost manner by the local institution according to the seriousness and context of the offence. The global science-based platform to discuss natural resources management is the World Resources Forum, based in Switzerland.

Land

Land, sometimes referred to as dry land, is the solid surface of the Earth that is not permanently covered by water. The vast majority of human activity occurs in land areas that support agriculture, habitat, and various natural resources.

Some life forms (including terrestrial plants and terrestrial animals) have developed from predecessor species that lived in bodies of water to exist on land.

Areas where land meets large bodies of water are called coastal zones. The division between land and water is a fundamental concept to humans, and can have strong cultural importance. The demarcation between land and water varies by local jurisdiction. A maritime boundary is one such political demarcation. A variety of natural boundaries exist to help define where water meets land. Solid rock landforms are easier to demarcate than marshy or swampy boundaries, where there is no clear point at which the land ends and a body of water has begun. Demarcation can further vary due to tides and weather.

History of Land on Earth

The earliest material found in the Solar System is dated to 4.5672±0.0006 bya (billion years ago); therefore, the Earth itself must have been formed by accretion around this time. By 4.54±0.04 bya, the primordial Earth had formed. The formation and evolution of the Solar System bodies occurred in tandem with the Sun. In theory a solar nebula partitions a volume out of a molecular cloud by gravitational collapse, which begins to spin and flatten into a circumstellar disk, which the planets then grow out of in tandem with the star. A nebula contains gas, ice grains and dust (including primordial nuclides). In nebular theory planetesimals commence forming as particulate accrues by cohesive clumping and then by gravity. The assembly of the primordial Earth proceeded for 10–20 myr.

Earth's atmosphere and oceans formed by volcanic activity and outgassing that included water vapor. The origin of the world's oceans was condensation augmented by water and ice delivered by asteroids, proto-planets, and comets. In this model, atmospheric "greenhouse gases" kept the oceans from freezing while the newly forming Sun was only at 70% luminosity. By 3.5 bya, the Earth's magnetic field was established, which helped prevent the atmosphere from being stripped away by the solar wind.

The crust, parts of which currently form the Earth's land, formed when the molten outer layer of the planet Earth cooled to form a solid as the accumulated water vapor began to act in the atmosphere. The two models that explain land mass propose either a steady growth to the present-day forms or, more likely, a rapid growth early in Earth history followed by a long-term steady continental area. Continents formed by plate tectonics, a process ultimately driven by the continuous loss of heat from the Earth's interior. On time scales lasting hundreds of millions of years, the supercontinents have formed and broken apart three times. Roughly 750 mya (million years ago), one of the earliest known supercontinents, Rodinia, began to break apart. The continents later recombined to form Pannotia, 600–540 mya, then finally Pangaea, which also broke apart 180 mya.

Land Mass

"Land mass" refers to the total surface area of the land of a geographical region or country (which may include discontinuous pieces of land such as islands). It is written as two words to distinguish it from the usage "landmass" —the contiguous area of land surrounded by ocean. The Earth's total land mass is 148,939,063.133 km^2 (57,505,693.767 sq mi) which is about 29.2% of its total surface. Water covers approximately 70.8% of the Earth's surface, mostly in the form of oceans.

Arable Land

Arable land (from Latin *arabilis*, "able to be plowed") is land *capable* of being ploughed and used to grow crops. In Britain, it was traditionally contrasted with pasturable lands such as heaths which could be used for sheep-rearing but not farmland.

In modern agriculture, however, as at the Food and Agriculture Organization, Eurostat, and the World Bank, "arable land" is a term of art meaning land that is *actually* being farmed (at minimum every five years) with crops that are sown and harvested within the same agricultural year. Arable land actually under crops in the present year is known as sown land or cropped land. The emended definition is preferred by the agencies because it distinguishes cultivable land that *could* be used to raise such annual crops but is instead devoted to "permanent cropland": for example, vinyards, orchards, and farms and plantations growing coffee, rubber, or nuts.

Arable Land Area

Although constrained by land mass and topography, the amount of arable land, both regionally and globally, fluctuates due to human and climatic factors such as irrigation, deforestation, desertification, terracing, land reclamation and urban sprawl. In 2008, the world's arable land amounted to 1,386 M ha, out of a total 4,883 M ha land used for agriculture.

Non-Arable Land

Land which is unsuitable for arable farming usually has at least one of the following deficiencies: no source of fresh water; too hot (desert); too cold (Arctic); too rocky; too mountainous; too salty; too rainy; too snowy; too polluted; or too nutrient poor. Clouds may block the sunlight plants need for photosynthesis, reducing productivity. Starvation and nomadism often exists on marginally arable land. Non-arable land is sometimes called *wasteland, badlands, worthless* or *no man's land.*

However, non-arable land can sometimes be converted into arable land. New arable land makes more food, and can reduce starvation. This outcome also makes a country more self-sufficient and politically independent, because food importation is reduced. Making non-arable land arable often involves digging new irrigation canals and new wells, aqueducts, desalination plants, planting trees for shade in the desert, hydroponics, fertilizer, nitrogen fertilizer, pesticides, reverse osmosis water processors, PET film insulation or other insulation against heat and cold, digging ditches and hills for protection against the wind, and greenhouses with internal light and heat for protection against the cold outside and to provide light in cloudy areas. This process is often extremely expensive. An alternative is the Seawater Greenhouse which desalinates water through evaporation and condensation using solar energy as the only energy input. This technology is optimized to grow crops on desert land close to the sea.

Some examples of infertile non-arable land being turned into fertile arable land are:

- *Aran Islands:* These islands off the west coast of Ireland, (not to be confused with the Isle of Arran in Scotland's Firth of Clyde), were unsuitable for arable farming because they were too rocky. The people covered the islands with a shallow layer of seaweed and sand from the ocean. This made it arable. Today, crops are grown there.
- *Israel:* Israel's land primarily consisted of desert until the construction of desalination plants along the country's coast. The desalination plants, which remove the salt from ocean water, have created a new source of water for farming, drinking, and washing.
- Slash and burn agriculture uses nutrients in wood ash, but these expire within a few years.
- Terra preta, fertile tropical soils created by adding charcoal.

Some examples of fertile arable land being turned into infertile land are:

- Droughts like the 'dust bowl' of the Great Depression in the U.S. turned farmland into desert.
- *Rainforest Deforestation:* The fertile tropical forests are converted into infertile desert land. For example, Madagascar's central highland plateau has become virtually totally barren (about ten percent of the country), as a result of slash-and-burn deforestation, an element of shifting cultivation practiced by many natives.
- Each year, arable land is lost due to desertification and human-induced erosion. Improper irrigation of farm land can wick the sodium, calcium, and magnesium from the soil and water to the surface. This process steadily concentrates salt in the root zone, decreasing productivity for crops that are not salt-tolerant.

Land Degradation

Land degradation is a process in which the value of the biophysical environment is affected by a combination of human-induced processes acting upon the land. Also environmental degradation is the gradual destruction or reduction of the quality and quantity of human activities animals activities or natural means example water causes soil erosion, wind, etc. It is viewed as any change or disturbance to the land perceived to be deleterious or undesirable. Natural hazards are excluded as a cause; however human activities can indirectly affect phenomena such as floods and bush fires.

This is considered to be an important topic of the 21st century due to the implications land degradation has upon agronomic productivity, the environment, and its effects on food security. It is estimated that up to 40% of the world's agricultural land is seriously degraded.

Measures

Land degradation is a broad term that can be applied differently across a wide range of scenarios. There are four main ways of looking at land degradation and its impact on the environment around it:

- A temporary or permanent decline in the productive capacity of the land. This can be seen through a loss of biomass, a loss of actual productivity or in potential productivity, or a loss or change in vegetative cover and soil nutrients.
- Action in the lands capacity to provide resources for human livelihoods. This can be measured from a base line of past land use.
- Loss of biodiversity: A loss of range of species or ecosystem complexity as a decline in the environmental quality.
- Shifting ecological risk: increased vulnerability of the environment or people to destruction or crisis. This is measured through a base line in the form of pre-existing risk of crisis or destruction.

A problem with defining land degradation is that what one group of people might view as degradation, others might view as a benefit or opportunity. For example, planting crops at a location with heavy rainfall and steep slopes would create scientific and environmental concern regarding the risk of soil erosion by water, yet farmers could view the location as a favourable one for high crop yields.

Different Types

In addition to the usual types of land degradation that have been known for centuries (water, wind and mechanical erosion, physical, chemical and biological degradation), four other types have emerged in the last 50 years:

- pollution, often chemical, due to agricultural, industrial, mining or commercial activities;
- loss of arable land due to urban construction;
- artificial radioactivity, sometimes accidental;
- land-use constraints associated with armed conflicts.

Overall, 36 types of land degradation can be assessed. All are induced or aggravated by human activities, e.g. sheet erosion, silting, aridification, salinization, urbanization, etc.

Causes

Figure: *Overgrazing by livestock can lead to land degradation*

Land degradation is a global problem, largely related to agricultural use. The major causes include:

- Land clearance, such as clearcutting and deforestation
- Agricultural depletion of soil nutrients through poor farming practices
- Livestock including overgrazing and overdrafting
- Inappropriate irrigation and overdrafting
- Urban sprawl and commercial development
- Soil contamination
- Vehicle off-roading
- Quarrying of stone, sand, ore and minerals
- Increase in field size due to economies of scale, reducing shelter for wildlife, as hedgerows and copses disappear
- Exposure of naked soil after harvesting by heavy equipment
- Monoculture, destabilizing the local ecosystem
- Dumping of non-biodegradable trash, such as plastics

Effects

Overcutting of vegetation occurs when people cut forests, woodlands and shrublands—to obtain timber, fuelwood and other products—at a pace exceeding the rate of natural regrowth. This is frequent in semi-arid environments, where fuelwood shortages are often severe.

Overgrazing is the grazing of natural pastures at stocking intensities above the livestock carrying capacity; the resulting decrease in the vegetation cover is a leading cause of wind and water erosion. It is a significant factor in Afghanistan. Next of land shortage the growing population pressure, during 1980-1990, has led to decreases in the already small areas of agricultural land per person in six out of eight countries (14% for India and 22% for Pakistan).

Population pressure also operates through other mechanisms. Improper agricultural practices, for instance, occur only under constraints such as the saturation of good lands under population pressure which leads settlers to cultivate too shallow or too steep soils, plough fallow land before it has recovered its fertility, or attempt to obtain multiple crops by irrigating unsuitable soils. High population density is not always related to land degradation. Rather, it is the practices of the human population that can cause a landscape to become degraded. Populations can be a benefit to the land and make it more productive than it is in its natural state. Land degradation is an important factor of internal displacement in many African and Asian countries.

Severe land degradation affects a significant portion of the Earth's arable lands, decreasing the wealth and economic development of nations. As the land resource base becomes less productive, food security is compromised and competition for dwindling resources increases, the seeds of famine and potential conflict are sown.

Sensitivity and Resilience

Sensitivity and resilience are measures of the vulnerability of a landscape to degradation. These two factors combine to explain the degree of vulnerability. Sensitivity is the degree to which a land system undergoes change due to natural forces, human intervention or a combination of both. Resilience is the ability of a landscape to absorb change, without significantly altering the relationship between the relative importance and numbers of individuals and species that compose the community. It also refers to the ability of the region to return to its original state after being changed in some way. The resilience of a landscape can be increased or decreased through human interaction based upon different methods of land-use management. Land that is degraded becomes less resilient than undergraded land, which can lead to even further degration through shocks to the landscape.

Climate Change

Significant land degradation from seawater inundation, particularly in river deltas and on low-lying islands, is a potential hazard that was identified in a 2007 IPCC report. As a result of sea-level rise from climate change, salinity levels can reach levels where agriculture becomes impossible in very low lying areas.

Chapter 2

Land Management

Land management is process of managing the use and development (in both urban and rural settings) of land resources. Land resources are used for a variety of purposes which may include organic agriculture, reforestation, water resource management and eco-tourism projects.

Soil

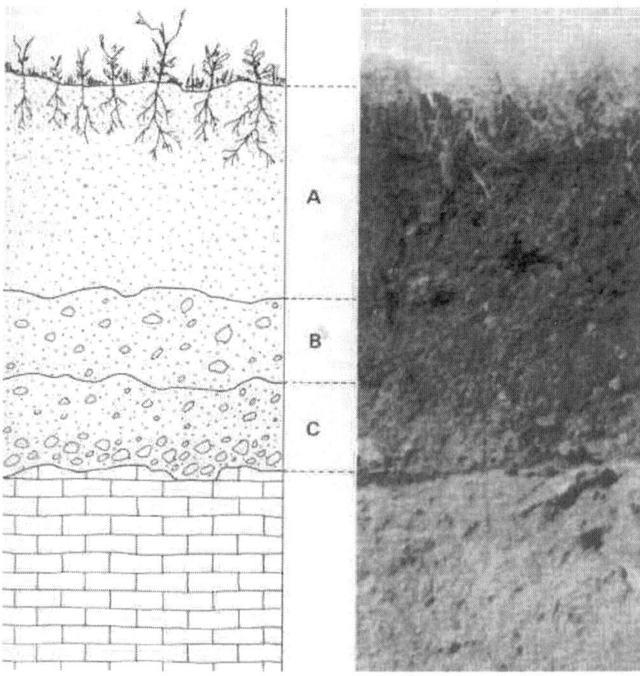

Figure: *A represents soil; B represents laterite, a regolith; C represents saprolite, a less-weathered regolith; the bottom-most layer represents bedrock*

Soil is the mixture of minerals, organic matter, gases, liquids, and the myriad of organisms that together support plant life. It is a natural body that exists as part of the pedosphere and which performs four important functions: it is a medium for plant growth; it is a means of water storage, supply and purification; it is a modifier of the atmosphere; and it is a habitat for organisms that take part in decomposition of organic matter and the creation of a habitat for new organisms.

Soil is considered to be the "skin of the earth" with interfaces between the lithosphere, hydrosphere, atmosphere, and biosphere. Soil consists of a solid phase (minerals and organic matter) as well as a porous phase that holds gases and water. Accordingly, soils are often treated as a three-state system.

Soil is the end product of the influence of the climate, relief (elevation, orientation, and slope of terrain), biotic activities (organisms), and parent materials (original minerals) interacting over time. Soil continually undergoes development by way of numerous physical, chemical and biological processes, which include weathering with associated erosion.

Most soils have a density between 1 and 2 g/cm^3. Little of the soil of planet Earth is older than the Pleistocene and none is older than the Cenozoic, although fossilized soils are preserved from as far back as the Archean. Soil science has two main branches of study: Edaphology and Pedology. Pedology is focused on the formation, description (morphology), and classification of soils in their natural environment, whereas edaphology is concerned with the influence of soils on organisms. In engineering terms, soil is referred to as regolith, or loose rock material that lies above the 'solid geology'. Soil is commonly referred to as "earth" or "dirt"; technically, the term "dirt" should be restricted to displaced soil.

As soil resources serve as a basis for food security, the international community advocates for its sustainable and responsible use through different types of Soil Governance.

Overview

Soil is a major component of the Earth's ecosystem. From ozone depletion and global warming to rain forest destruction and water pollution, the world's ecosystems are impacted in far-reaching ways by the processes carried out in the soil. Soil is the largest surficial global carbon reservoir on Earth, and it is potentially one of the most reactive to human disturbance and climate change. As the planet warms, soils will add carbon dioxide to the atmosphere due to its increased biological activity at higher temperatures. Thus, soil carbon losses likely have a large positive feedback response to global warming.

Soil acts as an engineering medium, a habitat for soil organisms, a recycling system for nutrients and organic wastes, a regulator of water quality, a modifier of atmospheric composition, and a medium for plant growth. Since soil has a tremendous range of available niches and habitats, it contains most of the earth's genetic diversity. A gram of soil can contain billions of organisms, belonging to thousands of species. Soil has a mean prokaryotic density of roughly 10^{13} organisms per cubic metre, whereas the ocean has a mean prokaryotic density of roughly 10^{8} organisms per cubic metre. The carbon content stored in soil is eventually returned to the atmosphere through the process of respiration, which is carried out by heterotrophic organisms that feed upon the carbonaceous material in the soil. Since plant roots need oxygen, ventilation is an important characteristic of soil. This ventilation can be accomplished via networks of soil pores, which also absorb and hold rainwater making it readily available for plant uptake. Since plants require a nearly continuous supply of water, but most regions receive sporadic rainfall, the water-holding capacity of soils is vital for plant survival.

Soils can effectively remove impurities, kill disease agents, and degrade contaminants. Typically, soils maintain a net absorption of oxygen and methane, and undergo a net release of carbon dioxide and nitrous oxide. Soils offer plants physical support, air, water, temperature moderation, nutrients, and protection from toxins. Soils provide readily available nutrients to plants and animals by converting dead organic matter into various nutrient forms.

Soils supply plants with mineral nutrients held in place by the clay and humus content of the soil. For optimum plant growth, the generalized content of soil components by volume should be roughly 50% solids (45% mineral and 5% organic matter), and 50% voids of which half is occupied by water and half by gas. The percent soil mineral and organic content is typically treated as a constant, while the percent soil water and gas content is considered highly variable whereby a rise in one is simultaneously balanced by a reduction in the other. The pore space allows for the infiltration and movement of air and water, both of which are critical for life in soil. Compaction, a common problem with soils, reduces this space, preventing air and water from reaching the plant roots and soil organisms.

Given sufficient time, an undifferentiated soil will evolve a soil profile which consists of two or more layers, referred to as soil horizons, that differ in one or more properties such as in their texture, structure, density, porosity, consistency, temperature, colour, and reactivity. The horizons differ greatly in thickness and generally lack sharp boundaries. Soil profile development is dependent on the processes that form soils from their parent materials, the type of parent material, and the factors that control soil formation. The biological influences on soil properties are strongest near the surface, while the geochemical influences on soil properties increase with depth. Mature soil profiles in temperate climate regions typically include three basic master horizons: A, B and C. The solum normally includes the A and B horizons. The living component of the soil is largely confined to the solum. In the more hot, humid, climate of the tropics, a soil may have only a single horizon.

The soil texture is determined by the relative proportions of sand, silt, and clay in the soil. The addition of organic matter, water, gases and time causes the soil of a certain texture to develop into a larger soil structure called an aggregate. At that point a soil can be said to be developed, and can be described further in terms of colour, porosity, consistency, reaction etc.

Of all the factors influencing the evolution of soil, water is the most powerful due to its involvement in the solution, erosion, transportation, and deposition of the materials of which a soil is composed. The mixture of water and dissolved and suspended materials is called the soil solution. Since soil water is never pure water, but contains hundreds of dissolved organic and inorganic substances, it may be more accurately called the soil solution. Water is central to the solution, precipitation and leaching of minerals from the soil profile. Finally, water affects the type of vegetation that grows in a soil, which in turn affects the development of the soil profile.

The most influential factor in stabilizing soil fertility are the soil colloidal particles, clay and humus, which behave as repositories of nutrients and moisture and so act to buffer the variations of soil solution ions and moisture. The contribution of soil colloids to soil nutrition are out of proportion to their part of the soil. Colloids act to store nutrients that might otherwise be leached from the soil or to release those ions in response to changes of soil pH, and so, to make them available to plants. The greatest influence on plant nutrient availability is soil pH, which is a measure of the hydrogen ion (acid-forming) soil reactivity, and is in turn a function of the soil materials, precipitation level, and plant root behaviour. Soil pH strongly affects the availability of nutrients.

Most nutrients, with the exception of nitrogen, originate from minerals. Some nitrogen originates from rain, but most of the nitrogen available in soils is the result of nitrogen fixation by bacteria. The action of microbes on organic matter and minerals may be to free nutrients for use, sequester them, or cause their loss from the soil by their volatilisation to gases or their leaching from the soil. The nutrients may be stored on soil colloids, and live or dead organic matter, but they may not be accessible to plants due to extremes of pH.

The organic material of the soil has a powerful effect on its development, fertility, and available moisture. Following water and soil colloids, organic material is next in importance to soil's formation and fertility.

History of the Study of Soil

Studies Concerning Soil Fertility

The history of the study of soil is intimately tied to our urgent need to provide food for ourselves and forage for our animals. Throughout history, civilizations have prospered or declined as a function of the availability and productivity of their soils.

The Greek historian Xenophon (450–355 B.C.) is credited with being the first to expound upon the merits of green-manuring crops: "But then whatever weeds are upon the ground, being turned into earth, enrich the soil as much as dung."

Columella's "Husbandry," circa 60 A.D., advocated the use of lime and that clover and alfalfa (green manure) should be turned under, and was used by 15 generations (450 years) under the Roman Empire until its collapse. From the fall of Rome to the French Revolution, knowledge of soil and agriculture was passed on from parent to child and as a result, crop yields were low. During the European Dark Ages, Yahya Ibn_al-'Awwam's handbook, with its emphasis on irrigation, guided the people of North Africa, Spain and the Middle East; a translation of this work was finally carried to the southwest of the United States.

Experiments into what made plants grow first led to the idea that the ash left behind when plant matter was burned was the essential element but overlooked the role of nitrogen, which is not left on the ground after combustion. In about 1635, the Flemish chemist Jan Baptist van Helmont thought he had proved water to be the essential element from his famous five years' experiment with a willow tree grown with only the addition of rainwater. His conclusion came from the fact that the increase in the plant's weight had apparently been produced only by the addition of water, with no reduction in the soil's weight. John Woodward (d. 1728) experimented with various types of water ranging from clean to muddy and found muddy water the best, and so he concluded that earthy matter was the essential element. Others concluded it was humus in the soil that passed some essence to the growing plant. Still others held that the vital growth principal was something passed from dead plants or animals to the new plants. At the start of the 18th century, Jethro Tull demonstrated that it was beneficial to cultivate (stir) the soil, but his opinion that the stirring made the fine parts of soil available for plant absorption was erroneous.

As chemistry developed, it was applied to the investigation of soil fertility. The French chemist Antoine Lavoisier showed in about 1778 that plants and animals must "combust" oxygen internally to live and was able to deduce that most of the 165-pound weight of van Helmont's willow tree derived from air. It was the French agriculturalist Jean-Baptiste Boussingault who by means of experimentation obtained evidence showing that the main sources of carbon, hydrogen and oxygen for plants were the air and water. Justus von Liebig in his book *Organic Chemistry in its Applications to Agriculture and Physiology* (published 1840), asserted that the chemicals in plants must have come from the soil and air and that to maintain soil fertility, the used minerals must be replaced. Liebig nevertheless believed the nitrogen was supplied from the air. The enrichment of soil with guano by the Incas was rediscovered in 1802, by Alexander von Humboldt. This led to its mining and that of Chilean nitrate and to its application to soil in the United States and Europe after 1840.

The work of Liebig was a revolution for agriculture, and so other investigators started experimentation based on it. In England John Bennet Lawes and Joseph Henry Gilbert worked in the Rothamsted Experimental Station, founded by the former, and discovered that plants took nitrogen from the soil, and that salts needed to be in an available state to be absorbed by plants. Their investigations also produced the "superphosphate", consisting in the acid treatment of phosphate rock. This led to the invention and use of salts of potassium (K) and nitrogen (N) as fertilizers. Ammonia generated by the production of coke was recovered and used as fertiliser. Finally, the chemical basis of nutrients delivered to the soil in manure was understood and in the mid-19th century chemical fertilisers were applied. However, the dynamic interaction of soil and its life forms awaited discovery.

In 1856 J. T. Way discovered that ammonia contained in fertilisers was transformed into nitrates, and twenty years later R. W. Warington proved that this transformation was done by living organisms. In 1890

Sergei Winogradsky announced he had found the bacteria responsible for this transformation.

It was known that certain legumes could take up nitrogen from the air and fix it to the soil but it took the development of bacteriology towards the end of the 19th century to lead to an understanding of the role played in nitrogen fixation by bacteria. The symbiosis of bacteria and leguminous roots, and the fixation of nitrogen by the bacteria, were simultaneously discovered by German agronomist Hermann Hellriegel and Dutch microbiologist Martinus Beijerinck.

Crop rotation, mechanisation, chemical and natural fertilisers led to a doubling of wheat yields in Western Europe between 1800 and 1900.

Studies Concerning Soil Formation

The scientists who studied the soil in connection with agricultural practices had considered it mainly as a static substrate. However, soil is the result of evolution from more ancient geological materials. After studies of the improvement of the soil commenced, others began to study soil genesis and as a result also soil types and classifications.

In 1860, in Mississippi, Eugene W. Hilgard studied the relationship among rock material, climate, and vegetation, and the type of soils that were developed. He realised that the soils were dynamic, and considered soil types classification. Unfortunately his work was not continued. At the same time Vasily Dokuchaev (about 1870) was leading a team of soil scientists in Russia who conducted an extensive survey of soils, finding that similar basic rocks, climate and vegetation types lead to similar soil layering and types, and established the concepts for soil classifications. Due to the language barriers, the work of this team was not communicated to Western Europe until 1914 by a publication in German by K. D. Glinka, a member of the Russian team.

Curtis F. Marbut was influenced by the work of the Russian team, translated Glinka's publication into English, and as he was placed in charge of the U. S. National Cooperative Soil Survey, applied it to a national soil classification system.

Soil-forming Processes

Soil formation, or pedogenesis, is the combined effect of physical, chemical, biological and anthropogenic processes working on soil parent material. Soil is said to be formed when organic matter has accumulated and colloids are washed downward, leaving deposits of clay, humus, iron oxide, carbonate, and gypsum, producing a distinct layer called the B horizon. (This is a somewhat arbitrary definition as mixtures of sand, silt, clay and humus will support biological and agricultural activity before that time.) These constituents are moved from one level to another by water and animal activity. As a result, layers (horizons) form in the soil profile. The alteration and movement of materials within a soil causes the formation of distinctive soil horizons.

How soil formation proceeds is influenced by at least five classic factors that are intertwined in the evolution of a soil. They are: parent material, climate, topography (relief), organisms, and time. When reordered to climate, relief, organisms, parent material, and time, they form the acronym CROPT.

An example of the development of a soil would begin with the weathering of lava flow bedrock, which would produce the purely mineral-based parent material from which the soil texture forms. Soil development would proceed most rapidly from bare rock of recent flows in a warm climate, under heavy and frequent rainfall. Under such conditions, plants become established very quickly on basaltic lava, even though there is very little organic material. The plants are supported by the porous rock as it is filled with nutrient-bearing water that carries dissolved minerals from the rocks and guano. Crevasses and pockets, local topography of the rocks, would hold fine materials and harbour plant roots. The developing plant roots are associated with mycorrhizal fungi that assist in breaking up the porous lava, and by these means organic matter and a finer mineral soil accumulate with time.

Parent Material

The mineral material from which a soil forms is called parent material. Rock, whether its origin is igneous, sedimentary, or metamorphic, is the source of all soil mineral materials and the origin of all plant nutrients with the exceptions of nitrogen, hydrogen and carbon. As the parent material is chemically and physically weathered, transported, deposited and precipitated, it is transformed into a soil.

Typical soil parent mineral materials are:

- Quartz: SiO_2
- Calcite: $CaCO_3$
- Feldspar: $KAlSi_3O_8$
- Mica (biotite): $K(Mg,Fe)_3AlSi_3O_{10}(OH)_2$

Classification of Parent Material

Parent materials are classified according to how they came to be deposited. Residual materials are mineral materials that have weathered in place from primary bedrock. Transported materials are those that have been deposited by water, wind, ice or gravity. Cumulose material is organic matter that has grown and accumulates in place.

Residual soils are soils that develop from their underlying parent rocks and have the same general chemistry as those rocks. The soils found on mesas, plateaux, and plains are residual soils. In the United States as little as three percent of the soils are residual.

Most soils derive from transported materials that have been moved many miles by wind, water, ice and gravity.

- Aeolian processes (movement by wind) are capable of moving silt and fine sand many hundreds of miles, forming loess soils (60–90 percent silt), common in the Midwest of North America and in Central Asia. Clay is seldom moved by wind as it forms stable aggregates.
- Water-transported materials are classed as either alluvial, lacustrine, or marine. Alluvial materials are those moved and deposited by flowing water. Sedimentary deposits settled in lakes are called lacustrine. Lake Bonneville and many soils around the Great Lakes of the United States are examples. Marine deposits, such as soils along the Atlantic and Gulf Coasts and in the Imperial Valley of California of the United States, are the beds of ancient seas that have been revealed as the land uplifted.
- Ice moves parent material and makes deposits in the form of terminal and lateral moraines in the case of stationary glaciers. Retreating glaciers leave smoother ground moraines and in all cases, outwash plains are left as alluvial deposits are moved downstream from the glacier.
- Parent material moved by gravity is obvious at the base of steep slopes as talus cones and is called colluvial material.

Cumulose parent material is not moved but originates from deposited organic material. This includes peat and muck soils and results from preservation of plant residues by the low oxygen content of a high water table. While peat may form sterile soils, muck soils may be very fertile.

Weathering of Parent Material

The weathering of parent material takes the form of physical weathering (disintegrating), chemical weathering (decomposition) and chemical transformation. Generally, minerals that are formed under the high temperatures and pressures at great depths within the earth's mantle are less resistant to weathering, while minerals formed at low temperature and pressure environment of the surface are more resistant to weathering. Weathering is usually confined to the top few metres of geologic material, because physical, chemical, and biological stresses generally decrease with depth. Physical disintegration begins as rocks that have solidified deep in the earth are exposed to lower pressure near the surface and swell and become unstable. Chemical decomposition is a function of mineral solubility, the rate of which doubles with each 10°C rise in temperature, but is strongly dependent on water to effect chemical changes. Rocks that will decompose in a few years in tropical climates will remain unaltered for millennia in deserts. Structural changes are the result of hydration, oxidation, and reduction.

- Physical disintegration is the first stage in the transformation of parent material into soil. Temperature fluctuations cause expansion and contraction of the rock, splitting it along lines of weakness. Water may then enter the cracks and freeze and cause the physical splitting of material along a path toward the centre of the rock, while temperature gradients within the rock can cause exfoliation of "shells". Cycles of wetting and drying cause soil particles to be abraded to a finer size, as does the physical rubbing of material as it is moved by wind, water, and gravity. Water can deposit within rocks minerals that expand upon drying, thereby stressing the rock. Finally, organisms reduce parent material in size through the action of plant roots or digging on the part of animals.
- Chemical decomposition and structural changes result when minerals are made soluble by water or are changed in structure. The first three of the following list are solubility changes and the last three are structural changes.

1. The solution of salts in water results from the action of bipolar water on ionic salt compounds producing a solution of ions and water.
2. Hydrolysis is the transformation of minerals into polar molecules by the splitting of the intervening water. This results in soluble acid-base pairs. For example, the hydrolysis of orthoclase-feldspar transforms it to acid silicate clay and basic potassium hydroxide, both of which are more soluble.
3. In carbonation, the reaction of carbon dioxide in solution with water forms carbonic acid. Carbonic acid will transform calcite into more soluble calcium bicarbonate.

4. Hydration is the inclusion of water in a mineral structure, causing it to swell and leaving it more stressed and easily decomposed.
5. Oxidation of a mineral compound is the inclusion of oxygen in a mineral, causing it to increase its oxidation number and swell due to the relatively large size of oxygen, leaving it stressed and more easily attacked by water (hydrolysis) or carbonic acid (carbonation).
6. Reduction the opposite of oxidation, means the removal of oxygen, hence oxidation number of some part of the mineral is reduced, which occurs when oxygen is scarce. The reduction of minerals leaves them electrically unstable, more soluble and internally stressed and easily decomposed.

Of the above, hydrolysis and carbonation are the most effective.

Saprolite is a particular example of a residual soil formed from the transformation of granite, metamorphic and other types of bedrock into clay minerals. Often called "weathered granite", saprolite is the result of weathering processes that include: hydrolysis, chelation from organic compounds, hydration (the solution of minerals in water with resulting cation and anion pairs) and physical processes that include freezing and thawing. The mineralogical and chemical composition of the primary bedrock material, its physical features, including grain size and degree of consolidation, and the rate and type of weathering transform the parent material into a different mineral. The texture, pH and mineral constituents of saprolite are inherited from its parent material.

Climate

The principal climatic variables influencing soil formation are effective precipitation (i.e., precipitation minus evapotranspiration) and temperature, both of which affect the rates of chemical, physical, and biological processes. The temperature and moisture both influence the organic matter content of soil through their effects on the balance between plant growth and microbial decomposition. Climate is the dominant factor in soil formation, and soils show the distinctive characteristics of the climate zones in which they form. For every 10 °C rise in temperature, the rates of biochemical reactions more than double. Mineral precipitation and temperature are the primary climatic influences on soil formation. If warm temperatures and abundant water are present in the profile at the same time, the processes of weathering, leaching, and plant growth will be maximized. Humid climates favour the growth of trees. In contrast, grasses are the dominant native vegetation in subhumid and semiarid regions, while shrubs and brush of various kinds dominate in arid areas.

Water is essential for all the major chemical weathering reactions. To be effective in soil formation, water must penetrate the regolith. The seasonal rainfall distribution, evaporative demand, site topography, and soil permeability interact to determine how effectively precipitation can influence soil formation. The greater the depth of water penetration, the greater the depth of weathering of the soil and its development. Surplus water percolating through the soil profile transports soluble and suspended materials from the upper to the lower layers. It may also carry away soluble materials in the drainage waters. Thus, percolating water stimulates weathering reactions and helps differentiate soil horizons. Likewise, a deficiency of water is a major factor in determining the characteristics of soils of dry regions. Soluble salts are not leached from these soils, and in some cases they build up to levels that curtail plant growth. Soil profiles in arid and semi-arid regions are also apt to accumulate carbonates and certain types of expansive clays.

The direct influences of climate include:

- A shallow accumulation of lime in low rainfall areas as caliche
- Formation of acid soils in humid areas
- Erosion of soils on steep hillsides
- Deposition of eroded materials downstream
- Very intense chemical weathering, leaching, and erosion in warm and humid regions where soil does not freeze

Climate directly affects the rate of weathering and leaching. Soil is said to be formed when detectable layers of clays, organic colloids, carbonates, or soluble salts have been moved downward. Wind moves sand and smaller particles, especially in arid regions where there is little plant cover. The type and amount of precipitation influence soil formation by affecting the movement of ions and particles through the soil, and aid in the development of different soil profiles. Soil profiles are more distinct in wet and cool climates, where organic materials may accumulate, than in wet and warm climates, where organic materials are rapidly consumed. The effectiveness of water in weathering parent rock material depends on seasonal and daily temperature fluctuations. Cycles of freezing and thawing constitute an effective mechanism which breaks up rocks and other consolidated materials.

Climate also indirectly influences soil formation through the effects of vegetation cover and biological activity, which modify the rates of chemical reactions in the soil.

Topography

The topography, or relief, is characterized by the inclination (slope), elevation, and orientation of the terrain. Topography determines the rate of precipitation or runoff and rate of formation or erosion of the surface soil profile. The topographical setting may either hasten or retard the work of climatic forces.

Steep slopes encourage rapid soil loss by erosion and allow less rainfall to enter the soil before running off and hence, little mineral deposition in lower profiles. In semiarid regions, the lower effective rainfall on steeper slopes also results in less complete vegetative cover, so there is less plant contribution to soil formation. For all of these reasons, steep slopes prevent the formation of soil from getting very far ahead of soil destruction. Therefore, soils on steep terrain tend to have rather shallow, poorly developed profiles in comparison to soils on nearby, more level sites.

In swales and depressions where runoff water tends to concentrate, the regolith is usually more deeply weathered and soil profile development is more advanced. However, in the lowest landscape positions, water may saturate the regolith to such a degree that drainage and aeration are restricted. Here, the weathering of some minerals and the decomposition of organic matter are retarded, while the loss of iron and manganese is accelerated. In such low-lying topography, special profile features characteristic of wetland soils may develop. Depressions allow the accumulation of water, minerals and organic matter and in the extreme, the resulting soils will be saline marshes or peat bogs. Intermediate topography affords the best conditions for the formation of an agriculturally productive soil.

Organisms

Soil is the most abundant ecosystem on Earth, but the vast majority of organisms in soil are microbes, a great many of which have not been described. There may be a population limit of around one billion cells per gram of soil, but estimates of the number of species vary widely. Estimates range from over a million species per gram of soil to 50,000 per gram of soil. The total number of organisms and species can vary widely according to soil type, location, and depth.

Plants, animals, fungi, bacteria and humans affect soil formation. Animals, soil mesofauna and micro-organisms mix soils as they form burrows and pores, allowing moisture and gases to move about. In the same way, plant roots open channels in soils. Plants with deep taproots can penetrate many metres through the different soil layers to bring up nutrients from deeper in the profile. Plants with fibrous roots that spread out near the soil surface have roots that are easily decomposed, adding organic matter. Micro-organisms, including fungi and bacteria, effect chemical exchanges between roots and soil and act as a reserve of nutrients.

Humans impact soil formation by removing vegetation cover with erosion as the result. Their tillage also mixes the different soil layers, restarting the soil formation process as less weathered material is mixed with the more developed upper layers.

Earthworms, ants and termites mix the soil as they burrow, significantly affecting soil formation. Earthworms ingest soil particles and organic residues, enhancing the availability of plant nutrients in the material that passes through their bodies. They aerate and stir the soil and increase the stability of soil aggregates, thereby assuring ready infiltration of water. As they build mounds, some organisms might transport soil materials from one horizon to another.

In general, the mixing activities of animals, sometimes called pedoturbation, tends to undo or counteract the tendency of other soil-forming processes that create distinct horizons. Termites and ants may also retard soil profile development by denuding large areas of soil around their nests, leading to increased loss of soil by erosion. Large animals such as gophers, moles, and prairie dogs bore into the lower soil horizons, bringing materials to the surface. Their tunnels are often open to the surface, encouraging the movement of water and air into the subsurface layers. In localized areas, they enhance mixing of the lower and upper horizons by creating, and later refilling, underground tunnels. Old animal burrows in the lower horizons often become filled with soil material from the overlying A horizon, creating profile features known as crotovinas.

Vegetation impacts soils in numerous ways. It can prevent erosion caused by excessive rain that might result from surface runoff. Plants shade soils, keeping them cooler and slow evaporation of soil moisture, or conversely, by way of transpiration, plants can cause soils to lose moisture. Plants can form new chemicals that can break down minerals and improve the soil structure. The type and amount of vegetation depends on climate, topography, soil characteristics, and biological factors. Soil factors such as density, depth, chemistry, pH, temperature and moisture greatly affect the type of plants that can grow in a given location. Dead plants and fallen leaves and stems begin their decomposition on the surface. There, organisms feed on them and mix the organic material with the upper soil layers; these added organic compounds become part of the soil formation process.

Human activities widely influence soil formation. For example, it is believed that Native Americans regularly set fires to maintain several large areas of prairie grasslands in Indiana and Michigan. In more recent times, human destruction of natural vegetation and subsequent tillage of the soil for crop production has abruptly modified soil formation. Likewise, irrigating an arid region of soil drastically influences the soil-forming factors, as does adding fertilizer and lime to soils of low fertility.

Time

Time is a factor in the interactions of all the above. While a mixture of sand, silt and clay constitute the texture of a soil and the aggregation of those components produces peds, the development of a soil with a distinct B horizon marks the development of a soil. With time, soils will evolve features that depend on the interplay of the prior listed soil-forming factors. It takes decades to several thousand years for a soil to develop a profile. That time period depends strongly on climate, parent material, relief, and biotic activity. For example, recently deposited material from a flood exhibits no soil development as there has not been enough time for the material to form a structure that further defines soil. The original soil surface is buried, and the formation process must begin anew for this deposit. Over time the soil will develop a profile that depends on the intensities of biota and climate. While a soil can achieve relative stability of its properties for extended periods, the soil life cycle ultimately ends in soil conditions that leave it vulnerable to erosion. Despite the inevitability of soil retrogression and degradation, most soil cycles are long.

Soil-forming factors continue to affect soils during their existence, even on "stable" landscapes that are long-enduring, some for millions of years. Materials are deposited on top or are blown or washed from the surface. With additions, removals and alterations, soils are always subject to new conditions. Whether these are slow or rapid changes depends on climate, topography and biological activity.

Physical Properties of Soils

The physical properties of soils, in order of decreasing importance, are texture, structure, density, porosity, consistency, temperature, colour and resistivity. Soil texture is determined by the relative proportion of the three kinds of soil particles, called soil separates: sand, silt, and clay. At the next larger scale, soil structures called peds are created from the soil separates when iron oxides, carbonates, clay, silica and humus, coat particles and cause them to adhere into larger, relatively stable secondary structures.

Soil density, particularly bulk density, is a measure of soil compaction. Soil porosity consists of the void part of the soil volume and is occupied by gases or water. Soil consistency is the ability of soil to stick together. Soil temperature and colour are self-defining. Resistivity refers to the resistance to conduction of electric currents and affects the rate of corrosion of metal and concrete structures. These properties may vary through the depth of a soil profile. Most of these properties determine the aeration of the soil and the ability of water to infiltrate and to be held within the soil.

Generalized Influence of Soil Texture Separates on Some Properties/Behavior of Soils			
Property/behavior	***Sand***	***Silt***	***Clay***
Water-holding capacity	Low	Medium to high	High
Aeration	Good	Medium	Poor
Drainage rate	High	Slow to medium	Very slow
Soil organic matter level	Low	Medium to high	High to medium
Decomposition of organic matter	Rapid	Medium	Slow
Warm-up in spring	Rapid	Moderate	Slow
Compactability	Low	Medium	High
Susceptibility to wind erosion	Moderate (High if fine sand)	High	Low
Susceptibility to water erosion	Low (unless fine sand)	High	Low if aggregated, otherwise high
Shrink/Swell Potential	Very Low	Low	Moderate to very high
Sealing of ponds, dams, and landfills	Poor	Poor	Good
Suitability for tillage after rain	Good	Medium	Poor
Pollutant leaching potential	High	Medium	Low (unless cracked)
Ability to store plant nutrients	Poor	Medium to High	High
Resistance to pH change	Low	Medium	High

Texture

The mineral components of soil are sand, silt and clay, and their relative proportions determine a soil's texture. Properties that are influenced by soil texture, include porosity, permeability, infiltration, shrink-swell rate, water-holding capacity, and susceptibility to erosion. In the illustrated USDA textural classification triangle, the only soil in which neither sand, silt nor clay predominates is called "loam". While even pure sand, silt or clay may be considered a soil, from the perspective of food production a loam soil with a small amount of organic material is considered ideal. The mineral constituents of a loam soil might be 40% sand, 40% silt and the balance 20% clay by weight. Soil texture affects soil behaviour, in particular its retention capacity for nutrients and water.

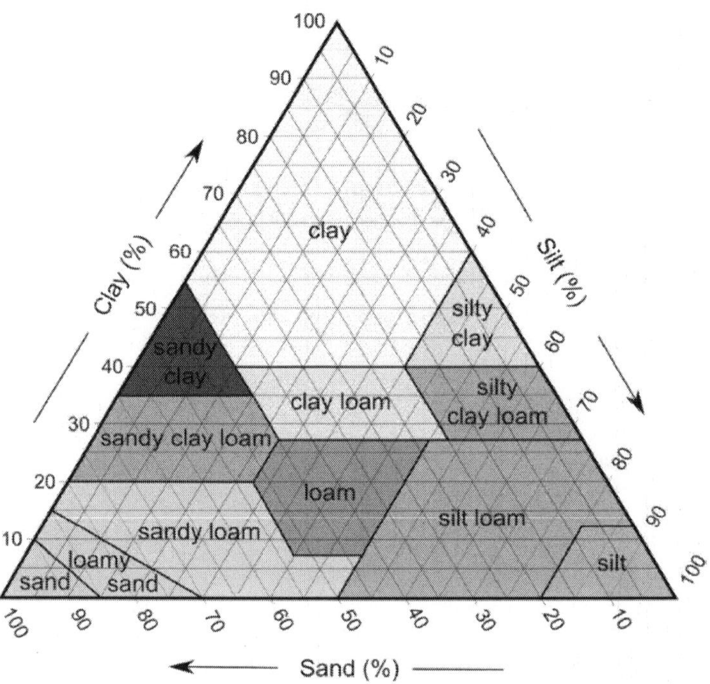

Figure: *Soil types by clay, silt and sand composition as used by the USDA*

Sand and silt are the products of physical and chemical weathering of the parent rock; clay, on the other hand, is a product of the precipitation of the dissolved parent rock as a secondary mineral. It is the large surface area to volume ratio (specific surface area) of soil particles and the unbalanced ionic charges within those that determine their role in the cation exchange capacity of soil, and hence its fertility. Sand is least active, followed by silt; clay is the most active. Sand's greatest benefit to soil is that it resists compaction and increases a soil's porosity. Silt is mineralogically like sand but with its higher specific surface area it is more chemically active than sand. But it is the clay content of soil, with its very high specific surface area and generally large number of negative charges, that gives a soil its high retention capacity for water and nutrients. Clay soils also resist wind and water erosion better than silty and sandy soils, as the particles bond tightly to each other.

Sand is the most stable of the mineral components of soil; it consists of rock fragments, primarily quartz particles, ranging in size from 2.0 to 0.05 mm (0.0787 to 0.0020 in) in diameter. Silt ranges in size from 0.05 to 0.002 mm (0.002 to 0.00008 in). Clay cannot be resolved by optical microscopes as its particles are 0.002 mm (7.9×10^{-5} in) or less in diameter. In medium-textured soils, clay is often washed downward through the soil profile and accumulates in the subsoil.

Soil components larger than 2.0 mm (0.079 in) are classed as rock and gravel and are removed before determining the percentages of the remaining components and the texture class of the soil, but are included in the name. For example, a sandy loam soil with 20% gravel would be called gravelly sandy loam.

When the organic component of a soil is substantial, the soil is called organic soil rather than mineral soil. A soil is called organic if:

1. Mineral fraction is 0% clay and organic matter is 20% or more
2. Mineral fraction is 0% to 50% clay and organic matter is between 20% and 30%
3. Mineral fraction is 50% or more clay and organic matter 30% or more.

Structure

The clumping of the soil textural components of sand, silt and clay causes aggregates to form and the further association of those aggregates into larger units creates soil structures called peds. The adhesion of the soil textural components by organic substances, iron oxides, carbonates, clays, and silica, and the breakage of those aggregates from expansion-contraction, freezing-thawing, and wetting-drying cycles, shape soil into distinct geometric forms. The peds evolve into units which may have various shapes, sizes and degrees of development. A soil clod, however, is not a ped but rather a mass of soil that results from mechanical disturbance of the soil. Soil structure affects aeration, water movement, conduction of heat, plant root growth and resistance to erosion. Water, in turn, has its strongest effect on soil structure due to its solution and precipitation of minerals and its effect on plant growth.

Soil structure often gives clues to its texture, organic matter content, biological activity, past soil evolution, human use, and the chemical and mineralogical conditions under which the soil formed. While texture is defined by the mineral component of a soil and is an innate property of the soil that does not change with agricultural activities, soil structure can be improved or destroyed by the choice and timing of farming practices.

Soil Structural Classes:

- *Types:* Shape and arrangement of peds
- *Platy:* Peds are flattened one atop the other 1-10 mm thick. Found in the A-horizon of forest soils and lake sedimentation.
- *Prismatic and Columnar:* Prismlike peds are long in the vertical dimension, 10-100 mm wide. Prismatic peds have flat tops, columnar peds have rounded tops. Tend to form in the B-horizon in high sodium soil where clay has accumulated.
- *Angular and subangular:* Blocky peds are imperfect cubes, 5-50 mm, angular have sharp edges, subangular have rounded edges. Tend to form in the B-horizon where clay has accumulated and indicate poor water penetration.
- *Granular and Crumb:* Spheroid peds of polyhedrons, 1-10 mm, often found in the A-horizon in the presence of organic material. Crumb peds are more porous and are considered ideal.
- *Classes:* Size of peds whose ranges depend upon the above type
- *Very fine or very thin:* <1 mm platy and spherical; <5 mm blocky; <10 mm prismlike.
- *Fine or thin:* 1-2 mm platy, and spherical; 5-10 mm blocky; 10-20 mm prismlike.
- *Medium:* 2-5 mm platy, granular; 10-20 mm blocky; 20-50 prismlike.
- *Coarse or thick:* 5-10 mm platy, granular; 20-50 mm blocky; 50-100 mm prismlike.
- *Very coarse or very thick:* >10 mm platy, granular; >50 mm blocky; >100 mm prismlike.
- *Grades:* Is a measure of the degree of development or cementation within the peds that results in their strength and stability.
- *Weak:* Weak cementation allows peds to fall apart into the three textural constituents, sand, silt and clay.
- *Moderate:* Peds are not distinct in undisturbed soil but when removed they break into aggregates, some broken aggregates and little unaggregated material. This is considered ideal.
- *Strong:* Peds are distinct before removed from the profile and do not break apart easily.
- *Structureless:* Soil is entirely cemented together in one great mass such as slabs of clay or no cementation at all such as with sand.

At the largest scale, the forces that shape a soil's structure result from swelling and shrinkage that initially tend to act horizontally, causing vertically oriented prismatic peds. Clayey soil, due to its differential drying rate with respect to the surface, will induce horizontal cracks, reducing columns to blocky peds. Roots, rodents, worms, and freezing-thawing cycles further break the peds into a spherical shape.

At a smaller scale, plant roots extend into voids and remove water causing the open spaces to increase, and decrease physical aggregation size. At the same time roots, fungal hyphae and earthworms create microscopic tunnels that break up peds.

At an even smaller scale, soil aggregation continues as bacteria and fungi exude sticky polysaccharides which bind soil into small peds. The addition of the raw organic matter that bacteria and fungi feed upon encourages the formation of this desirable soil structure.

At the lowest scale, the soil chemistry affects the aggregation or dispersal of soil particles. The clay particles contain polyvalent cations which give the faces of clay layers a net negative charge. At the same time the edges of the clay plates have a slight positive charge, thereby allowing the edges to adhere to the faces of other clay particles or to flocculate (form clumps). On the other hand, when monovalent ions such as sodium invade and displace the polyvalent cations, they weaken the positive charges on the edges, while the negative surface charges are relatively strengthened. This leaves a net negative charge on the clay, causing the particles to push apart, and by doing so to prevent the flocculation of clay particles into larger, open assemblages. As a result, the clay disperses and settles into voids between peds, causing those to close. In this way the soil aggregation is destroyed and the soil is made impenetrable to air and water. Such sodic soil tends to form columnar structures near the surface.

Density

Density is the weight per unit volume of an object. Particle density is equal to the mass of solid particles divided by the volume of solid particles - it is the density of only the mineral particles that make up a soil; i.e., it excludes pore space and organic material. Soil particle density is typically 2.60 to 2.75 grams per cm^3 and is usually unchanging for a given soil. Soil particle density is lower for soils with high organic matter content, and is higher for soils with high Fe-oxides content. Soil bulk density is equal to the dry mass of the soil divided by the volume of the soil; i.e., it includes air space and organic materials of the soil volume. A high bulk density is indicative of either soil compaction or high sand content. The bulk density of cultivated loam is about 1.1 to 1.4 g/cm^3 (for comparison water is 1.0 g/cm^3). Soil bulk density is highly variable for a given soil. A lower bulk density by itself does not indicate suitability for plant growth due to the influence of soil texture and structure. Soil bulk density is inherently always less than the soil particle density.

Representative bulk densities of soils. The percentage pore space was calculated using 2.7 g/cm^3 for particle density except for the peat soil, which is estimated.		
Soil treatment and identification	*Bulk density g/cm^3*	*Pore space %*
Tilled surface soil of a cotton field	1.3	51
Trafficked inter-rows where wheels passed surface	1.67	37
Traffic pan at 25 cm deep	1.7	36
Undisturbed soil below traffic pan, clay loam	1.5	43
Rocky silt loam soil under aspen forest	1.62	40
Loamy sand surface soil	1.5	43
Decomposed peat	0.55	65

Porosity

Pore space is that part of the bulk volume of soil that is not occupied by either mineral or organic matter but is open space occupied by either gases or water. Ideally, the total pore space should be 50% of the soil volume. The gas space is needed to supply oxygen to organisms decomposing organic matter, humus, and plant roots. Pore space also allows the movement and storage of water and dissolved nutrients. This property of soils effectively compartmentalizes the soil pore space such that many organisms are not in direct competition with one another, which may explain not only the large number of species present, but the fact that functionally redundant organisms (organisms with the same ecological niche) can co-exist within the same soil.

There are four categories of pores:

1. Very fine pores: < 2 μm
2. Fine pores: 2-20 μm
3. Medium pores: 20-200 μm
4. Coarse pores: 200 μm-0.2 mm

In comparison, root hairs are 8 to 12 μm in diameter. When pore space is less than 30 μm, the forces of attraction that hold water in place are greater than the gravitational force acting to drain the water. At that point, soil becomes water-logged and it cannot breathe. For a growing plant, pore size is of greater importance than total pore space. A medium-textured loam provides the ideal balance of pore sizes. Having large pore spaces that allow rapid gas and water movement is superior to smaller pore space soil that has a greater percentage pore space. Soil texture determines the pore space at the smallest scale, but at a larger scale, soil structure has a strong influence on soil aeration, water infiltration and drainage. Tillage has the short-term benefit of temporarily increasing the number of pores of largest size, but in the end those will be degraded by the destruction of soil aggregation. Clay soils have smaller pores, but more total pore space than sand.

Consistency

Consistency is the ability of soil to stick to itself or to other objects (cohesion and adhesion respectively) and its ability to resist deformation and rupture. It is of approximate use in predicting cultivation problems and the engineering of foundations. Consistency is measured at three moisture conditions: air-dry, moist, and wet; In those conditions the consistency quality depend upon the clay content. In the wet state, the two qualities of stickiness and plasticity are assessed. A soil's resistance to fragmentation and crumbling is assessed in the dry state by rubbing the sample. Its resistance to shearing forces is assessed in the moist state by thumb and finger pressure. Finally, a soil's plasticity is measured in the wet state by moulding with the hand. Finally, the cemented consistency depends on cementation by substances other than clay, such as calcium carbonate, silica, oxides and salts; moisture content has little effect on its assessment. The measures of consistency border on subjective as they employ the "feel" of the soil in those states.

The terms used to describe the soil consistency in three moisture states and a last consistency not affected by the amount of moisture are as follows:

1. Consistency of Dry Soil: loose, soft, slightly hard, hard, very hard, extremely hard
2. Consistency of Moist Soil: loose, very friable, friable, firm, very firm, extremely firm
3. Consistency of Wet Soil: nonsticky, slightly sticky, sticky, very sticky; nonplastic, slightly plastic, plastic, very plastic
4. Consistency of Cemented Soil: weakly cemented, strongly cemented, indurated (requires hammer blows to break up)

Soil consistency is useful in estimating the ability of soil to support buildings and roads. More precise measures of soil strength are often made prior to construction.

Temperature

Soil temperature depends on the ratio of the energy absorbed to that lost. Soil has a temperature range between -20 to 60 °C. Soil temperature regulates seed germination, plant and root growth and the availability of nutrients. Below 50 cm (20 in), soil temperature seldom changes and can be approximated by adding 1.8 °C (2 °F) to the mean annual air temperature. Soil temperature has important seasonal, monthly and daily variations. Fluctuations in soil temperature are much lower with increasing soil depth. Heavy mulching (a type of soil cover) can slow the warming of soil, and, at the same time, reduce fluctuations in surface temperature.

Most often, agricultural activities must adapt to soil temperatures by:

1. maximizing germination and growth by timing of planting
2. optimizing use of anhydrous ammonia by applying to soil below 10 °C (50 °F)
3. preventing heaving and thawing due to frosts from damaging shallow-rooted crops
4. preventing damage to desirable soil structure by freezing of saturated soils
5. improving uptake of phosphorus by plants

Otherwise soil temperatures can be raised by drying soils or the use of clear plastic mulches. Organic mulches slow the warming of the soil.

There are various factors that affect soil temperature, such as water content, soil colour, and relief (slope, orientation, and elevation), and soil cover (shading and insulation). The colour of the ground cover and its insulating properties have a strong influence on soil temperature. Whiter soil tends to have a higher albedo than blacker soil cover, which encourages whiter soils to have cooler soil temperatures.

The specific heat of soil is the energy required to raise the temperature of soil by 1 °C. The specific heat of soil increases as water content increases, since the heat capacity of water is greater than that of dry soil. The specific heat of pure water is ~ 1 calorie per gram, the specific heat of dry soil is ~ 0.2 calories per gram and the specific heat of wet soil is ~ 0.2 to 1 calories per gram. Also, tremendous energy (~540 cal/g) is required and dissipated to evaporate water (known as the heat of vaporization). As such, wet soil usually warms more slowly than dry soil - wet surface soil is typically 3 to 6 °C colder than dry surface soil.

Soil heat flux refers to the conduction (or movement) of energy (or heat) in response to a temperature gradient. The heat flux density is the amount of energy that flows through soil per unit area per unit time has both magnitude and direction.

$$\vec{q} = -k\nabla T$$

where (including the SI units)

$\vec{q}$ is the local heat flux, W·m^{-2}

k is the material's conductivity, W·m^{-1}·K^{-1},

∇T where the del operator applied to the temperature gives the temperature gradient, K·m^{-1} and has both magnitude and direction.

The thermal conductivity, k, is often treated as a constant, though this is not always true. While the thermal conductivity of a material generally varies with temperature, the variation is generally small over a significant range of temperatures for some common materials. In anisotropic materials, the thermal conductivity typically varies with orientation; in this case is represented by a second-order tensor. In nonuniform materials, varies with spatial location. For soil, thermal conductivity also depends on mineral composition, water content, and bulk density. Compact and wet soils have a higher thermal conductivity than loose and dry soils. For many simple applications, Fourier's law is used in its one-dimensional, x-direction form:

Component	***Thermal Conductivity (W.m 1.K 1)***
Quartz	8.8
Clay	2.9
Organic Matter	0.25
Water	0.57
Ice	2.4
Air	0.025
Dry soil	0.2 0.4
Wet soil	1-3

Colour

Soil colour is often the first impression one has when viewing soil. Striking colours and contrasting patterns are especially noticeable. The Red River (Mississippi watershed) carries sediment eroded from extensive reddish soils like Port Silt Loam in Oklahoma. The Yellow River in China carries yellow sediment from eroding loess soils. Mollisols in the Great Plains of North America are darkened and enriched by organic matter. Podsols in boreal forests have highly contrasting layers due to acidity and leaching.

In general, colour is determined by the organic matter content, drainage conditions, and degree of oxidation. Soil colour, while easily discerned, has little use in predicting soil characteristics." Arizona Master Gardener Manual". Cooperative Extension, College of

Agriculture, University of Arizona. p. Chapter 2, pp 4–8. Retrieved 27 May 2013. It is of use in distinguishing boundaries within a soil profile, determining the origin of a soil's parent material, as an indication of wetness and waterlogged conditions, and as a qualitative means of measuring organic, salt and carbonate contents of soils. Colour is recorded in the Munsell colour system as for instance 10YR3/4.

Soil colour is primarily influenced by soil mineralogy. Many soil colours are due to various iron minerals. The development and distribution of colour in a soil profile result from chemical and biological weathering, especially redox reactions. As the primary minerals in soil parent material weather, the elements combine into new and colourful compounds. Iron forms secondary minerals of a yellow or red colour, organic matter decomposes into black and brown compounds, and manganese, sulfur and nitrogen can form black mineral deposits. These pigments can produce various colour patterns within a soil. Aerobic conditions produce uniform or gradual colour changes, while reducing environments (anaerobic) result in rapid colour flow with complex, mottled patterns and points of colour concentration.

Resistivity

Soil resistivity is a measure of a soil's ability to retard the conduction of an electric current. The electrical resistivity of soil can affect the rate of galvanic corrosion of metallic structures in contact with the soil. Higher moisture content or increased electrolyte concentration can lower resistivity and increase conductivity, thereby increasing the rate of corrosion. Soil resistivity values typically range from about 2 to 1000 $\Omega \cdot m$, but more extreme values are not unusual.

Soil Water

Water affects soil formation, structure, stability and erosion but is of primary concern with respect to plant growth. Water is essential to plants for four reasons:

1. It constitutes 80%-95% of the plant's protoplasm.
2. It is essential for photosynthesis.
3. It is the solvent in which nutrients are carried to, into and throughout the plant.
4. It provides the turgidity by which the plant keeps itself in proper position.

In addition, water alters the soil profile by dissolving and re-depositing minerals, often at lower levels, and possibly leaving the soil sterile in the case of extreme rainfall and drainage. In a loam soil, solids constitute half the volume, gas one-quarter of the volume, and water one-quarter of the volume of which only half will be available to most plants.

A flooded field will drain the gravitational water under the influence of gravity until water's adhesive and cohesive forces resist further drainage at which point it has finally reached field capacity. At that point, plants must apply suction to draw water from a soil. When soil becomes too dry, the available water is used up and the remaining moisture is unavailable water as the plant cannot produce sufficient suction to draw the water in. A plant must produce suction that increases from zero for a flooded field to 1/3 bar at field dry condition. At 15 bar suction, wilting percent, seeds will not germinate, plants begin to wilt and then die. Water moves in soil under the influence of gravity, osmosis and capillarity. When water enters the soil, it displaces air from some of the pores, since air content of a soil is inversely related to its water content.

The rate at which a soil can absorb water depends on the soil and its other conditions. As a plant grows, its roots remove water from the largest pores first. Soon the larger pores hold only air, and the remaining water is found only in the intermediate- and smallest-sized pores. The water in the smallest pores is so strongly held on particle surfaces that plant roots cannot pull it away. Consequently, not all soil water is available to plants. When saturated, the soil may lose nutrients as the water drains. Water moves in a drained field under the influence of pressure where the soil is locally saturated and by capillarity pull. Most plant water needs are supplied from the suction caused by of evaporation from plant leaves and 10% is supplied by "suction" created by osmotic pressure differences between the plant interior and the soil water. Plant roots must seek out water. Insufficient water will damage the yield of a crop. Most of the available water is used in transpiration to pull nutrients into the plant.

Water Retention Forces

Water is retained in a soil when the adhesive force of attraction of water's hydrogen atoms for the oxygen of soil particles and the cohesive forces water's hydrogen feels for other water's oxygen atoms are stronger than the forces that might pull it from the soil. When a field is flooded, the air in the soil pore space is displaced by water. The field will drain under the force of gravity until it reaches what is called field capacity, at which point the smallest pores are filled with water and the largest with water and gases. The total amount of water held when field capacity is reached is a function of the specific surface area of the soil particles. As a result, high clay and high organic soils have higher field capacities. The total force required to pull or push water out of soil is termed suction and usually expressed in units of bars (10^5 pascal) which is just a little less than one-atmosphere pressure. Alternatively, the terms "tension" or "moisture potential" may be used.

Moisture Classification

The forces with which water is held in soils determine its availability to plants. Forces of adhesion hold water strongly to mineral and humus surfaces and less strongly to itself by cohesive forces. A plant's root may penetrate a very small volume of water that is adhering to soil and be initially able to draw in water that is only lightly held by the cohesive forces. But as the droplet is drawn down, the forces of adhesion of the water for the soil particles make reducing the volume of water increasingly difficult until the plant cannot produce sufficient suction to use the remaining water. The remaining water is considered unavailable. The amount of available water depends upon the soil texture and humus amounts and the type of plant attempting to use the water. Cacti, for example, can produce greater suction than can agricultural crop plants.

The following description applies to a loam soil and agricultural crops. When a field is flooded, it is said to be saturated and all available air space is occupied by water. The suction required to draw water into a plant root is zero. As the field drains under the influence of gravity (drained water is called gravitational water or drain-able water), the suction a plant must produce to use such water increases to 1/3 bar. At that point, the soil is said to have reached field capacity, and plants that use the water must produce increasingly higher suction, finally up to 15 bar. At 15 bar suction, the soil water amount is called wilting percent. At that suction the plant cannot sustain its water needs as water is still being lost from the plant by transpiration; the plant's turgidity is lost, and it wilts. The next level, called air-dry, occurs at 1000 bar suction. Finally the oven dry condition is reached at 10,000 bar suction. All water below wilting percentage is called unavailable water.

Soil Moisture Content

When the soil moisture content is optimal for plant growth, the water in the large and intermediate size pores can move about in the soil and be easily used by plants. The amount of water remaining in a soil drained to field capacity and the amount that is available are functions of the soil type. Sandy soil will retain very little water, while clay will hold the maximum amount. The time required to drain a field from flooded condition for a clay loam that begins at 43% water by weight to a field capacity of 21.5% is six days, whereas a sand loam that is flooded to its maximum of 22% water will take two days to reach field capacity of 11.3% water. The available water for the clay loam might be 11.3% whereas for the sand loam it might be only 7.9% by weight.

Wilting point, field capacity, and available water capacity of various soil textures

Wilting point, field capacity, and available water capacity of various soil textures						
Soil Texture	***Wilting Point***		***Field Capacity***		***Available water capacity***	
	Water per foot of soil depth		***Water per foot of soil depth***		***Water per foot of soil depth***	
	%	in.	%	in.	%	in.
Medium sand	1.7	0.3	6.8	1.2	5.1	0.9
Fine sand	2.3	0.4	8.5	1.5	6.2	1.1
Sandy loam	3.4	0.6	11.3	2.0	7.9	1.4
Fine sandy loam	4.5	0.8	14.7	2.6	10.2	1.8
Loam	6.8	1.2	18.1	3.2	11.3	2.0
Silt loam	7.9	1.4	19.8	3.5	11.9	2.1
Clay loam	10.2	1.8	21.5	3.8	11.3	2.0
Clay	14.7	2.6	22.6	4.0	7.9	1.4

The above are average values for the soil textures as the percentage of sand, silt and clay vary within the listed soil textures.

Water Flow in Soils

Water moves through soil due to the force of gravity, osmosis and capillarity. At zero to one-third bar suction, water is pushed through soil from the point of its application under the force of gravity and the pressure gradient created by the pressure of the water; this is called saturated flow. At higher suction, water movement is pulled by capillarity from wetter toward dryer soil. This is caused by water's adhesion to soil solids, and is called unsaturated flow.

Water infiltration and movement in soil is controlled by six factors:

1. Soil texture
2. Soil structure. Fine-textured soils with granular structure are most favourable to infiltration of water.
3. The amount of organic matter. Coarse matter is best and if on the surface helps prevent the destruction of soil structure and the creation of crusts.
4. Depth of soil to impervious layers such as hardpans or bedrock
5. The amount of water already in the soil
6. Soil temperature. Warm soils take in water faster while frozen soils may not be able to absorb depending on the type of freezing.

Water infiltration rates range from 0.25 cm (0.098 in) per hour for high clay soils to 2.5 cm (0.98 in) per hour for sand and well stabilised and aggregated soil structures. Water flows through the ground unevenly, called "gravity fingers", because of the surface tension between water particles. Tree roots create paths for rainwater flow through soil by breaking though soil including clay layers: one study showed roots increasing infiltration of water by 153% and another study showed an increase by 27 times. Flooding temporarily increases soil permeability in river beds, helping to recharge aquifers.

Saturated Flow

Water applied to a soil is pushed by pressure gradients from the point of its application where it is saturated locally to less saturated areas. Once soil is completely wetted, any more water will move downward, or percolate, carrying with it clay, humus and nutrients, primarily cations, out of the range of plant roots. In order of decreasing solubility, the leached nutrients are:

- Calcium
- Magnesium, Sulfur, Potassium; depending upon soil composition
- Nitrogen; usually little, unless nitrate fertiliser was applied recently
- Phosphorus; very little as its forms in soil are of low solubility.

In the United States percolation water due to rainfall ranges from zero inches just east of the Rocky Mountains to twenty or more inches in the Appalachian Mountains and the north coast of the Gulf of Mexico.

Unsaturated Flow

At suctions less than one-third bar, water moves in all directions via unsaturated flow at a rate that is dependent on the square of the diameter of the water-filled pores. Water is pulled by capillary action due to the adhesion force of water to the soil solids, producing a suction gradient from wet towards drier soil. Doubling the diameter of the pores increases the flow rate by a factor of four. Large pores drained by gravity and not filled with water do not greatly increase the flow rate for unsaturated flow. Water flow is primarily from coarse-textured soil into fine-textured soil and is slowest in fine-textured soils such as clay.

Water Uptake by Plants

Of equal importance to the storage and movement of water in soil is the means by which plants acquire it and their nutrients. Ninety percent of water is taken up by plants as passive absorption caused by the pulling force of water evaporating (transpiring) from the long column of water that leads from the plant's roots to its leaves. In addition, the high concentration of salts within plant roots creates an osmotic pressure gradient that pushes soil water into the roots. Osmotic absorption becomes more important during times of low water transpiration caused by lower temperatures (for example at night) or high humidity. It is the process that causes guttation.

Root extension is vital for plant survival. A study of a single winter rye plant grown for four months in one cubic foot of loam soil showed that the plant developed 13,800,000 roots a total of 385 miles in length and 2,550 square feet in surface area and 14 billion hair roots of 6,600 miles total length and 4,320 square feet total area, for a total surface area of 6,870 square feet (83 ft squared). The total surface area of the loam soil was estimated to be 560,000 square feet. In other words the roots were in contact with only 1.2% of the soil. Roots must seek out water as the unsaturated flow of water in soil can move only at a rate of up to 2.5 cm (one inch) per day; as a result they are constantly dying and growing as they seek out high concentrations of soil moisture.

Insufficient soil moisture, to the point of causing wilting, will cause permanent damage and crop yields will suffer. When grain sorghum was exposed to soil suction as low as 13.0 bar during the seed head emergence through bloom and seed set stages of growth, its production was reduced by 34%.

Consumptive use and Water Efficiency

Only a small fraction (0.1% to 1%) of the water used by a plant is held within the plant. The majority is ultimately lost via transpiration, while evaporation from the soil surface is also substantial. Transpiration plus evaporative soil moisture loss is called evapotranspiration. Evapotranspiration plus water held in the plant totals to consumptive use, which is nearly identical to evapotranspiration.

The total water used in an agricultural field includes runoff, drainage and consumptive use. The use of loose mulches will reduce evaporative losses for a period after a field is irrigated, but in the end the total evaporative loss will approach that of an uncovered soil. The benefit from mulch is to keep the moisture available during the seedling stage. Water use efficiency is measured by transpiration ratio, which is the ratio of the total water transpired by a plant to the dry weight of the harvested plant. Transpiration ratios for crops range from 300 to 700. For example alfalfa may have a transpiration ratio of 500 and as a result 500 kilograms of water will produce one kilogram of dry alfalfa.

Soil Atmosphere

The atmosphere of soil is radically different from the atmosphere above. The consumption of oxygen, by microbes and plant roots and their release of carbon dioxide, decrease oxygen and increase carbon dioxide concentration. Atmospheric CO_2 concentration is 0.04%, but in the soil pore space it may range from 10 to 100 times that level. At extreme levels CO_2 is toxic. In addition, the soil voids are saturated with water vapour. Adequate porosity is necessary not just to allow the penetration of water but also to allow gases to diffuse in and out. Movement of gases is by diffusion from high concentrations to lower. Oxygen diffuses in and is consumed and excess levels of carbon dioxide, diffuse out with other gases as well as water. Soil texture and structure strongly affect soil porosity and gas diffusion. It is the total pore space (porosity) of soil not the pore size that determines the rate of diffusion of gases into and out of soil. A Platy soil structure and compacted soils (low porosity) impede gas flow, and a deficiency of oxygen may encourage anaerobic bacteria to reduce nitrate NO_3 to the gases N_2, N_2O, and NO, which are then lost to the atmosphere. Aerated soil is also a net sink of methane CH_4 but a net producer of greenhouse gases when soils are depleted of oxygen and subject to elevated temperatures.

Composition of Soil Particles

Soil particles can be classified by their chemical composition (mineralogy) as well as their size. The particle size distribution of a soil, its texture, determines many of the properties of that soil, but the mineralogy of those particles can strongly modify those properties. The mineralogy of the finest soil particles, clay, is especially important.

Gravel, Sand and Silt

Gravel, sand and silt are the larger soil particles, and their mineralogy is often inherited from the parent material of the soil, but may include products of weathering (such as concretions of calcium carbonate or iron oxide), or residues of plant and animal life (such as silica phytoliths). Quartz is the most common mineral in the sand or silt fraction as it is resistant to chemical weathering; other common minerals are felspars, micas and ferromagnesian minerals such as pyroxenes, amphiboles and olivines.

Mineral Colloids; Soil Clays

Due to its high specific surface area and its unbalanced negative charges, clay is the most active mineral component of soil. It is a colloidal and most often a crystalline material. In soils, clay is defined in a physical sense as any mineral particle less than 2 μm (8×10^{-5} in) in effective diameter. Chemically, clay is a range of minerals with certain reactive properties. Clay is also a soil textural class. Many soil minerals, such as gypsum, carbonates, or quartz, are small enough to be classified physically as clay but chemically they do not afford the same utility as do clay minerals.

Clay was once thought to be very small particles of quartz, feldspar, mica, hornblende or augite, but it is now known to be (with the exception of mica-based clays) a precipitate with a mineralogical composition that is dependent on but different from its parent materials and is classed as a secondary mineral. The type of clay that is formed is a function of the parent material and the composition of the minerals in solution. Clay minerals continue to be formed as long as the soil exists. Mica-based clays result from a modification of the primary mica mineral in such a way that it behaves and is classed as a clay. Most clays are crystalline, but some are amorphous. The clays of a soil are a mixture of the various types of clay, but one type predominates.

Most clays are crystalline and most are made up of three or four planes of oxygen held together by planes of aluminium and silicon by way of ionic bonds that together form a single layer of clay. The spatial arrangement of the oxygen atoms determines clay's structure. Half of the weight of clay is oxygen, but on a volume basis oxygen is ninety percent. The layers of clay are sometimes held together through hydrogen bonds or potassium bridges and as a result will swell less in the presence of water. Other clays, such as montmorillonite, have layers that are loosely attached and will swell greatly when water intervenes between the layers.

There are three groups of clays:

1. Crystalline alumino-silica clays: montmorillonite, illite, vermiculite, chlorite, kaolinite.
2. Amorphous clays: young mixtures of silica (SiO_2-OH) and alumina ($Al(OH)_3$) which have not had time to form regular crystals.
3. Sesquioxide clays: old, highly leached clays which result in oxides of iron, aluminium and titanium.

Alumino-silica Clays

Alumino-silica clays are characterised by their regular crystalline structure. Oxygen in ionic bonds with silicon forms a tetrahedral coordination (silicon at the centre) which in turn forms sheets of silica. Two sheets of silica are bonded together by a plane of aluminium which forms an octahedral coordination, called alumina, with the oxygens of the silica sheet above and that below it. Hydroxyl ions (OH^-) sometimes substitute for oxygen. During the clay formation process, Al^{3+} may substitute for Si^{4+}, and as much as one fourth of the aluminium Al^{3+} may be substituted by Zn^{2+}, Mg^{2+} or Fe^{2+}. The substitution of lower-valence cations for higher-valence cations (isomorphic substitution) gives clay a net negative charge that attracts and holds soil cations, some of which are of value for plant growth. Isomorphic substitution occurs during the clay's formation and does not change with time.

- Montmorillonite clay is made of four planes of oxygen with two silicon and one central aluminium plane intervening. The alumino-silicate montmorillonite clay is said to have a 2:1 ratio of silicon to aluminium. The seven planes together form a single layer of montmorillonite. The layers are weakly held together and water may intervene, causing the clay to swell up to ten times its dry volume. It occurs in soils which have had little leaching, hence it is found in arid regions. The entire surface is exposed and available for surface reactions and it has a high cation exchange capacity (CEC).
- Illite is a 2:1 clay similar in structure to montmorillonite but has potassium bridges between the clay layers and the degree of swelling depends on the degree of weathering of the potassium. The active surface area is reduced due to the potassium bonds. Illite originates from the modification of mica, a primary mineral. It is often found together with montmorillonite and its primary minerals. It has moderate CEC.
- Vermiculite is a mica-based clay similar to illite, but the layers of clay are held together more loosely by hydrated magnesium and it will swell, but not as much as does montmorillonite. It has very high CEC.
- Chlorite is similar to vermiculite, but the loose bonding by occasional hydrated magnesium, as in vermiculite, is replaced by a hydrated magnesium sheet, that firmly bonds the planes above and below it. It has two planes of silicon, one of aluminium and one of magnesium; hence it is a 2:2 clay. Chlorite does not swell and it has low CEC.
- Kaolinite is very common, more common than montmorillonite in acid soils. It has one silica and one alumina sheet per layer; hence it is a 1:1 type clay. One layer of oxygen is replaced with hydroxyls, which produces strong hydrogen bonds to the oxygen in the next layer of clay. As a result kaolinite does not swell in

water and has a low specific surface area, and as almost no isomorphic substitution has occurred it has a low CEC. Where rainfall is high, acid soils selectively leach more silica than alumina from the original clays, leaving kaolinite. Even heavier weathering results in sesquioxide clays.

Amorphous Clays

Amorphous clays are young, and commonly found in volcanic ash. They are mixtures of alumina and silica which have not formed the ordered crystal shape of alumino-silica clays which time would provide. The majority of their negative charges originates from hydroxyl ions, which can gain or lose a hydrogen ion (H^+) in response to soil pH, in such way was as to buffer the soil pH. They may have either a negative charge provided by the attached hydroxyl ion (OH^-), which can attract a cation, or lose the hydrogen of the hydroxyl to solution and display a positive charge which can attract anions. As a result they may display either high CEC, in an acid soil solution, or high anion exchange capacity, in a basic soil solution.

Sesquioxide Clays

Sesquioxide clays are a product of heavy rainfall that has leached most of the silica and alumina from alumino-silica clay, leaving the less soluble oxides of iron Fe_2O_3 and iron hydroxide ($Fe(OH)_3$) and aluminium hydroxides ($Al(OH)_3$). It takes hundreds of thousands of years of leaching to create sesquioxide clays. Sesqui is Latin for "one and one-half": there are three parts oxygen to two parts iron or aluminium; hence the ratio is one and one-half. They are hydrated and act as either amorphous or crystalline. They are not sticky and do not swell, and soils high in them behave much like sand and can rapidly pass water. They are able to hold large quantities of phosphates. Sesquioxides have low CEC. Such soils range from yellow to red in colour. Such clays tend to hold phosphorus tightly rendering them unavailable for absorption by plants.

Organic Colloids

Humus is the penultimate state of decomposition of organic matter; while it may linger for a thousand years, on the larger scale of the age of the mineral soil components, it is temporary. It is composed of the very stable lignins (30%) and complex sugars (polyuronides, 30%), proteins (30%), waxes, and fats that are resistant to breakdown by microbes. Its chemical assay is 60% carbon, 5% nitrogen, some oxygen and the remainder hydrogen, sulfur, and phosphorus. On a dry weight basis, the CEC of humus is many times greater than that of clay.

Carbon and Terra Preta

In the extreme environment of high temperatures and the leaching caused by the heavy rain of tropical rain forests, the clay and organic colloids are largely destroyed. The heavy rains wash the aluminosilicate clays from the soil leaving only sesquioxide clays of low CEC. The high temperatures and humidity allow bacteria and fungi to virtually dissolve any organic matter on the rain-forest floor overnight and much of the nutrients are volatilized or leached from the soil and lost. However, carbon in the form of charcoal is far more stable than soil colloids and is capable of performing many of the functions of the soil colloids of sub-tropical soils. Soil containing substantial quantities of charcoal, of an anthropogenic origin, is called terra preta. Research into terra preta is still young but is promising. Fallow periods "on the Amazonian Dark Earths can be as short as 6 months, whereas fallow periods on oxisols are usually 8 to 10 years long"

Soil Chemistry

The chemistry of a soil determines its ability to supply available plant nutrients and affects its physical properties and the health of its microbial population. In addition, a soil's chemistry also determines its corrosivity, stability, and ability to absorb pollutants and to filter water. It is the surface chemistry of mineral and organic colloids that determines soil's chemical properties. "A colloid is a small, insoluble, nondiffusible particle larger than a molecule but small enough to remain suspended in a fluid medium without settling. Most soils contain organic colloidal particles called humus as well as the inorganic colloidal particles of clays." The very high specific surface area of colloids and their net charges, gives soil its great ability to hold and release ions. Negatively charged sites on colloids attract and release cations in what is referred to as cation exchange. Cation-exchange capacity (CEC) is the amount of exchangeable cations per unit weight of dry soil and is expressed in terms of milliequivalents of positively charged ions per 100 grams of soil (or centimoles of positive charge per kilogram of soil; $cmol_c/kg$). Similarly, positively charged sites on colloids can attract and release anions in the soil giving the soil anion exchange capacity (AEC).

Cation and Anion Exchange

The cation exchange, that takes place between colloids and soil water, buffers (moderates) soil pH, alters soil structure, and purifies percolating water by adsorbing cations of all types, both useful and harmful.

The negative or positive charges on colloid particles make them able to hold cations or anions, respectively, to their surfaces. The charges result from four sources.

- Isomorphous substitution occurs in clay when lower-valence cations substitute for higher-valence cations in the crystal structure. Substitutions in the outermost layers are more effective than for the innermost layers, as the charge strength drops off as the square of the distance. The net result is a negative charge.
- Edge-of-clay oxygen atoms are not in balance ionically as the tetrahedral and octahedral structures are incomplete.
- Hydroxyls may substitute for oxygens of the silica layers. When the hydrogens of the clay hydroxyls are ionised into solution, they leave the oxygen with a negative charge.
- Hydrogens of humus hydroxyl groups may be ionised into solution, leaving an oxygen with a negative charge.

Cations held to the negatively charged colloids resist being washed downward by water and out of reach of plants' roots, thereby preserving the fertility of soils in areas of moderate rainfall and low temperatures.

There is a hierarchy in the process of cation exchange on colloids, as they differ in the strength of adsorption by the colloid and hence their ability to replace one another. If present in equal amounts in the soil water solution:

Al^{3+} replaces H^{+} replaces Ca^{2+} replaces Mg^{2+} replaces K^{+} same as NH^{4+} replaces Na^{+}

If one cation is added in large amounts, it may replace the others by the sheer force of its numbers. This is called mass action. This is largely what occurs with the addition of fertiliser.

As the soil solution becomes more acidic (an abundance of H^{+}), the other cations more weakly bound to colloids are pushed into solution and hydrogen ions occupy those sites. A low pH may cause hydrogen of hydroxyl groups to be pulled into solution, leaving charged sites on the colloid available to be occupied by other cations. This ionisation of hydroxyl groups on the surface of soil colloids creates what is described as pH-dependent charges. Unlike permanent charges developed by isomorphous substitution, pH-dependent charges are variable and increase with increasing pH. Freed cations can be made available to plants but are also prone to be leached from the soil, possibly making the soil less fertile. Plants are able to excrete H^{+} into the soil and by that means, change the pH of the soil near the root and push cations off the colloids, thus making those available to the plant.

Cation Exchange Capacity (CEC)

Cation exchange capacity should be thought of as the soil's ability to remove cations from the soil water solution and sequester those to be exchanged later as the plant roots release hydrogen ions to the solution. CEC is the amount of exchangeable hydrogen cation (H^+) that will combine with 100 grams dry weight of soil and whose measure is one milliequivalents per 100 grams of soil (1 meq/100 g). Hydrogen ions have a single charge and one-thousandth of a gram of hydrogen ions per 100 grams dry soil gives a measure of one milliequivalent of hydrogen ion. Calcium, with an atomic weight 40 times that of hydrogen and with a valence of two, converts to (40/2) x 1 milliequivalent = 20 milliequivalents of hydrogen ion per 100 grams of dry soil or 20 meq/ 100 g. The modern measure of CEC is expressed as centimoles of positive charge per kilogram (cmol/kg) of oven-dry soil.

Most of the soil's CEC occurs on clay and humus colloids, and the lack of those in hot, humid, wet climates, due to leaching and decomposition respectively, explains the relative sterility of tropical soils. Live plant roots also have some CEC.

Cation exchange capacity for soils; soil textures; soil colloids		
Soil	***State***	***CEC meq/100 g***
Charlotte fine sand	Florida	1.0
Ruston fine sandy loam	Texas	1.9
Glouchester loam	New Jersey	11.9
Grundy silt loam	Illinois	26.3
Gleason clay loam	California	31.6
Susquehanna clay loam	Alabama	34.3
Davie mucky fine sand	Florida	100.8
Sands	------	1 - 5
Fine sandy loams	------	5-10
Loams and silt loams	-----	5-15
Clay loams	-----	15-30
Clays	-----	over 30
Sesquioxides	-----	0-3
Kaolinite	-----	3-15
Illite	-----	25-40
Montmorillonite	-----	60-100
Vermiculite (similar to illite)	-----	80-150
Humus	-----	100-300

Anion Exchange Capacity (AEC)

Anion exchange capacity should be thought of as the soil's ability to remove anions from the soil water solution and sequester those for later exchange as the plant roots release carbonate anions to the soil water solution. Those colloids which have low CEC tend to have some AEC. Amorphous and sesquioxide clays have the highest AEC, followed by the iron oxides. Levels of AEC are much lower than for CEC. Phosphates tend to be held at anion exchange sites.

Iron and aluminum hydroxide clays are able to exchange their hydroxide anions (OH^-) for other anions. The order reflecting the strength of anion adhesion is as follows:

$H_2PO_4^-$ replaces SO_4^{2-} replaces NO_3^- replaces Cl^-

The amount of exchangeable anions is of a magnitude of tenths to a few milliequivalents per 100 g dry soil. As pH rises, there are relatively more hydroxyls, which will displace anions from the colloids and force them into solution and out of storage; hence AEC decreases with increasing pH (alkalinity).

Soil Reaction (pH)

Soil reactivity is expressed in terms of pH and is a measure of the acidity or alkalinity of the soil. More precisely, it is a measure of hydrogen ion concentration in an aqueous solution and ranges in values from 0 to 14 (acidic to basic) but practically speaking for soils, pH ranges from 3.5 to 9.5, as pH values beyond those extremes are toxic to life forms.

Soil pH

At 25°C an aqueous solution that has a pH of 3.5 has $10^{-3.5}$ moles H^+ (hydrogen ions) per litre of solution (and also $10^{-10.5}$ mole/litre OH^-). A pH of 7, defined as neutral, has 10^{-7} moles hydrogen ions per litre of solution and also 10^{-7} moles of OH^- per litre; since the two concentrations are equal, they are said to neutralise each other. A pH of 9.5 has $10^{-9.5}$ moles hydrogen ions per litre of solution (and also $10^{-2.5}$ mole per litre OH^-). A pH of 3.5 has one million times more hydrogen ions per litre than a solution with pH of 9.5 (9.5 - 3.5 = 6 or 10^6) and is more acidic.

The effect of pH on a soil is to remove from the soil or to make available certain ions. Soils with high acidity tend to have toxic amounts of aluminium and manganese. Plants which need calcium need moderate alkalinity, but most minerals are more soluble in acid soils. Soil organisms are hindered by high acidity, and most agricultural crops do best with mineral soils of pH 6.5 and organic soils of pH 5.5.

In high rainfall areas, soils tend to acidity as the basic cations are forced off the soil colloids by the mass action of hydrogen ions from the rain as those attach to the colloids. High rainfall rates can then wash the nutrients out, leaving the soil sterile. Once the colloids are saturated with H^+, the addition of any more hydrogen ions or aluminum hydroxyl cations drives the pH even lower (more acidic) as the soil has been left with no buffering capacity. In areas of extreme rainfall and high temperatures, the clay and humus may be washed out, further reducing the buffering capacity of the soil. In low rainfall areas, unleached calcium pushes pH to 8.5 and with the addition of exchangeable sodium, soils may reach pH 10. Beyond a pH of 9, plant growth is reduced. High pH results in low micro-nutrient mobility, but water-soluble chelates of those nutrients can supply the deficit. Sodium can be reduced by the addition of gypsum (calcium sulphate) as calcium adheres to clay more tightly than does sodium causing sodium to be pushed into the soil water solution where it can be washed out by an abundance of water.

Base Saturation Percentage

There are acid-forming cations (hydrogen and aluminium) and there are base-forming cations. The fraction of the base-forming cations that occupy positions on the soil colloids is called the base saturation percentage. If a soil has a CEC of 20 meq and 5 meq are aluminium and hydrogen cations (acid-forming), the remainder of positions on the colloids (20-5 = 15 meq) are assumed occupied by base-forming cations, so that the percentage base saturation is 15/20 x 100% = 75% (the compliment 25% is assumed acid-forming cations). When the soil pH is 7 (neutral), base saturation is 100 percent and there are no hydrogen ions stored on the colloids. Base saturation is almost in direct proportion to pH (increases with increasing pH). It is of use in calculating the amount of lime needed to neutralise an acid soil. The amount of lime needed to neutralize a soil must take account of the amount of acid forming ions on the colloids not just those in the soil water solution. The addition of enough lime to neutralize the soil water solution will be insufficient to change the pH, as the acid forming cations stored on the soil colloids will tend to restore the original pH condition as they are pushed off those colloids by the calcium of the added lime.

Buffering of Soils

The resistance of soil to changes in pH as a result of the addition of acid or basic material is a measure of the buffering capacity of a soil and increases as the CEC increases. Hence, pure sand has almost no buffering ability, while soils high in colloids have high buffering capacity. Buffering occurs by cation exchange and neutralisation.

The addition of a small amount highly basic aqueous ammonia to a soil will cause the ammonium to displace hydrogen ions from the colloids, and the end product is water and colloidally fixed ammonium, but no permanent change overall in soil pH.

The addition of a small amount of lime, $CaCO_3$, will displace hydrogen ions from the soil colloids, causing the fixation of calcium to colloids and the evolution of CO_2 and water, with no permanent change in soil pH.

The addition of carbonic acid (the solution of CO_2 in water) will displace calcium from colloids, as hydrogen ions are fixed to the colloids, evolving water and slightly alkaline (temporary increase in pH) highly soluble calcium bicarbonate, which will then precipitate as lime ($CaCO_3$) and water at a lower level in the soil profile, with the result of no permanent change in soil pH.

All of the above are examples of the buffering of soil pH. The general principal is that an increase in a particular cation in the soil water solution will cause that cation to be fixed to colloids (buffered) and a decrease in solution of that cation will cause it to be withdrawn from the colloid and moved into solution (buffered). The degree of buffering is limited by the CEC of the soil; the greater the CEC, the greater the buffering capacity of the soil.

Nutrients

Sixteen nutrients are essential for plant growth and reproduction. They are carbon, hydrogen, oxygen, nitrogen, phosphorus, potassium, sulfur, calcium, magnesium, iron, boron, manganese, copper, zinc, molybdenum, and chlorine. Nutrients required for plants to complete their life cycle are considered essential nutrients. Nutrients that enhance the growth of plants but are not necessary to complete the plant's life cycle are considered non-essential. With the exception of carbon, hydrogen and oxygen, which are supplied by carbon dioxide and water, the nutrients derive originally from the mineral component of the soil. Although minerals are the origin of those nutrients, the organic component of the soil is the reservoir of the majority of readily available plant nutrients. For the nutrients to be available to plants, they must be in the proper ionic form (with the exception of water and CO_2). For example, the application of finely ground minerals, feldspar and apatite, to soil does not provide the necessary amounts of potassium and phosphorus for good plant growth. Nitrogen is the primary limiting nutrient and phosphorus is second to nitrogen as the primary nutrient for plants, animals and microorganisms.

The provision of plant nutrition involves chemical, biological, and physical processes. Nearly all plant nutrients are taken up from the soil water solution in the form of ions, either cations or as anions. In an effort to gain nutrients, plants will release ions to the soil. Bicarbonate (HCO_3^-) and hydroxyl (OH^-) anions released from plant roots enhance the absorption of nutrient anions; similarly, hydrogen cations are released in exchange for cation forms of nutrients. As a result, nutrient ions are pushed into the soil water solution from their sequestration on colloids to become available to plants. Nitrogen, for example, is available in soil organic material but is unusable by plants until it is made available by that material's decomposition by micro-organisms into cation or anion forms. The NH_4^+ (ammonium) and NO_3^- (nitrate) forms of nitrogen are stored on the soil colloids until forced off those by the presence of other cations and anions. After that, they will move by physical means to near the plant roots. Generally, plant roots can readily absorb all of the nutrients from the soil solution, provided there is enough oxygen gas in the soil to support root metabolism.

The bulk of most nutrient elements in the soil is held in the form of primary and secondary minerals, and organic matter. The primary minerals (mostly rock dust in the form of silt) hold the nutrients too tightly to be readily available; the nutrients adsorbed onto the colloids clay and humus are moderately available; and the soil solution fraction has ions that are freely available for absorption by plant roots. Gram for gram, the capacity of humus to hold nutrients and water is far greater than that of clay. All in all, small amounts of humus may remarkably increase the soil's capacity to promote plant growth.

Table: *Plant nutrients, their chemical symbols, and the ionic forms common in soils and available for plant uptake*

Element	***Symbol***	***Ion or molecule***
Carbon	C	CO_2 (mostly through leaves)
Hydrogen	H	H^+, HOH (water)
Oxygen	O	O^{2-}, OH^-, CO_3^{2-}, SO_4^{2-}, CO_2
Phosphorus	P	$H_2PO_4^-$, HPO_4^{2-} (phosphates)
Potassium	K	K^+
Nitrogen	N	NH_4^+, NO_3^- (ammonium, nitrate)
Sulfur	S	SO_4^{2-}
Calcium	Ca	Ca^{2+}
Iron	Fe	Fe^{2+}, Fe^{3+} (ferrous, ferric)
Magnesium	Mg	Mg^{2+}

Contd...

Element	***Symbol***	***Ion or molecule***
Boron	B	H_3BO_3, $H_2BO_3^-$, $B(OH)_4^-$
Manganese	Mn	Mn^{2+}
Copper	Cu	Cu^{2+}
Zinc	Zn	Zn^{2+}
Molybdenum	Mo	MoO_4^{2-} (molybdate)
Chlorine	Cl	Cl^- (chloride)

Mechanism of Nutrient Uptake

All the nutrients with the exception of carbon are taken up by the plant through its roots. To be taken up by a plant, a nutrient element must be in an ionic form (with the exception of water and H_3BO_3) and must be located at the root surface. Often, parts of a root are in such intimate contact with soil particles that a direct exchange may take place between nutrient ions adsorbed on the surface of the soil colloids and hydrogen ions from the surface of root cell walls. In any case, the supply of nutrients in contact with the root would soon be depleted. There are however, three basic mechanisms by which the concentration of nutrient ions at the root surface is maintained. Nutrient levels are maintained by three principal mechanisms by which nutrient ions dissolved in the soil solution are brought into contact with plant roots:

1. Mass flow
2. Diffusion
3. Root interception

All three mechanisms operate simultaneously, but one mechanism or another may be most important for a particular nutrient. For example, in the case of calcium, which is generally plentiful in the soil solution, mass flow alone can usually bring sufficient amounts to the root surface. However, in the case of phosphorus, diffusion is needed to supplement mass flow because the soil solution is very low in this element in comparison to the amounts needed by plants. For the most part, nutrient ions must travel some distance in the soil solution to reach the root surface. This movement can take place by mass flow, as when dissolved nutrients are carried along with the soil water flowing toward a root that is actively drawing water from the soil. In this type of movement, the nutrient ions are somewhat analogous to leaves floating down a stream. In addition, nutrient ions continually move by diffusion from areas of greater concentration toward the nutrient-depleted areas of lower concentration around the root surface. By this means, plants can continue to take up nutrients even at night, when water is only slowly absorbed into the roots. Finally, root interception comes into play as roots continually grow into new, undepleted soil.

Because nutrient uptake is an active metabolic process, conditions that inhibit root metabolism may also inhibit nutrient uptake. Examples of such conditions include excessive soil water content or soil compaction resulting in poor soil aeration, excessively high or low soil temperatures, and above-ground conditions that result in low translocation of sugars to plant roots. A maize plant will use one quart of water per day at the height of its growing season.

Estimated relative importance of mass flow, diffusion and root interception as mechanisms in supplying plant nutrients to corn plant roots in soils			
Nutrient	*Approximate percentage supplied by:*		
	Mass flow	*Root interception*	*Diffusion*
Nitrogen	98.8	1.2	0
Phosphorus	6.3	2.8	90.9
Potassium	20.0	2.3	77.7
Calcium	71.4	28.6	0
Sulfur	95.0	5.0	0
Molybdenum	95.2	4.8	0

In the above table, phosphorus and potassium nutrients move more by diffusion than they do by mass flow in water solution, as they are rapidly taken up by the roots creating a concentration of almost zero near the roots. The very steep concentration gradient is of greater influence in the movement of those ions than is the movement of those by mass flow. The movement by mass flow requires the transpiration of water from the plant causing water and solution ions to also move toward the roots. Movement by root interception is slowest as the plants must extend their roots. Plants move ions out of their roots in an effort to move nutrients in from the soil. Hydrogen H^+ is exchanged for cations, and carbonate (HCO_3^-) and hydroxide (OH^-) anions are exchanged for nutrient anions. Plants derive most of their anion nutrients from decomposing organic matter, which holds 95 percent of the nitrogen, 5 to 60 percent of the phosphorus and 80 percent of the sulfur. As plant roots remove nutrients from the soil water solution, nutrients are added to the soil water as other ions move off of clay and humus, are added from the decomposition of soil minerals, and are released by the decomposition of organic matter. Where crops are produced, the replenishment of nutrients in the soil must be augmented by the addition of fertiliser or organic matter.

Carbon

Plants obtain their carbon from atmospheric carbon dioxide. A plant's weight is forty-five percent carbon. Elementally, carbon is 50%

of plant material. Plant residues have a carbon to nitrogen ratio (C/N) of 50:1. As the soil organic material is digested by arthropods and micro-organisms, the C/N decreases as the carbonaceous material is metabolised and carbon dioxide (CO_2) is released as a byproduct which then finds its way out of the soil and into the atmosphere. The nitrogen, and other nutrients however, is sequestered in the bodies of the living matter of those organisms and so it builds up in the soil. Normal CO_2 concentration in the atmosphere is 0.03%, which is probably the factor limiting plant growth. In a field of maize on a still day during high light conditions in the growing season, the CO_2 concentration drops very low, but under such conditions the crop could use up to 20 times the normal concentration. The respiration of CO_2 by soil micro-organisms decomposing soil organic matter contributes an important amount of CO_2 to the photosynthesising plants. Within the soil, CO_2 concentration is 10 to 100 times that of atmospheric levels but may rise to toxic levels if the soil porosity is low or if diffusion is impeded by flooding.

Nitrogen

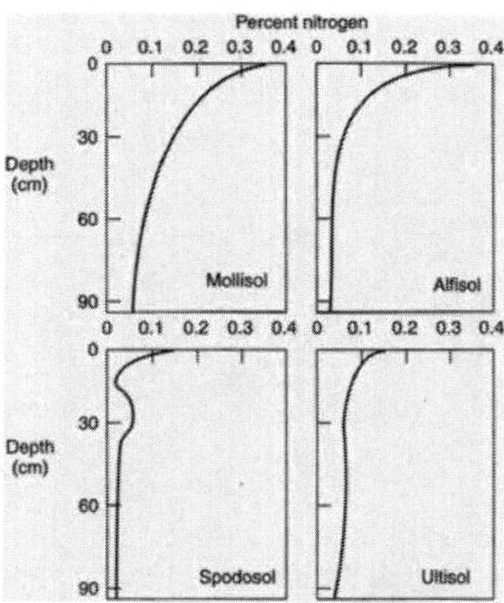

Figure: *Generalization of percent soil nitrogen by soil order*

Nitrogen is the most critical element obtained by plants from the soil and is a bottleneck in plant growth. Plants can use the nitrogen as either the ammonium cation (NH_4^+) or the anion nitrate (NO_3^-). Nitrogen is seldom missing in the soil, but is often in the form of raw organic material which cannot be used directly. The total nitrogen content depends on the climate, vegetation, topography, age and soil management. Soil nitrogen typically decreases by 0.2 to 0.3% for every temperature increase by 10 °C. Usually, more nitrogen is under grassland than under forest. Humus formation promotes nitrogen immobilization. Cultivation decreases soil nitrogen by exposing soil to more air which the bacteria can use, and no-tillage maintains more nitrogen than tillage.

***Table:** Carbon/Nitrogen Ratio of Various Organic Materials*

Organic Material	*C:N Ratio*
Alfalfa	13
Bacteria	4
Clover, green sweet	16
Clover, mature sweet	23
Fungi	9
Forest litter	30
Humus in warm cultivated soils	11
Legume-grass hay	25
Legumes (alfalfa or clover), mature	20
Manure, cow	18
Manure, horse	16–45
Manure, human	10
Oat straw	80
Straw, cornstalks	90
Sawdust	250

Some micro-organisms are able to metabolise organic matter and release ammonium in a process called *mineralisation.* Others take free ammonium and oxidise it to nitrate. Particular bacteria are capable of metabolising N_2 into the form of nitrate in a process called nitrogen fixation. Both ammonium and nitrate can be *immobilized* or essentially lost from the soil by its incorporation into the microbes' living cells, where it is temporarily sequestered in the form of amino acids and protein. Nitrate may also be lost from the soil when bacteria metabolise it to the gases N_2 and N_2O. The loss of gaseous forms of nitrogen to the atmosphere due to microbial action is called *denitrification.* Nitrogen may also be *leached* from the soil if it is in the form of nitrate or lost to the atmosphere as ammonia due to a chemical reaction of ammonium with alkaline soil by way of a process called *volatilisation.* Ammonium may also be sequestered in clay by *fixation.* A small amount of nitrogen is added to soil by rainfall.

Nitrogen Gains

In the process of mineralisation, microbes feed on organic matter, releasing ammonia (NH_3) (which may be reduced to ammonium (NH_4^+)

and other nutrients. As long as the carbon to nitrogen ratio (C/N) in the soil is above 30:1, nitrogen will be in short supply and other bacteria will feed on the ammonium and incorporate its nitrogen into their cells in the immobilization process. In that form the nitrogen is said to be immobilised. Later, when such bacteria die, they too are mineralised and some of the nitrogen is released as ammonium and nitrate. If the C/N is less than 15, ammonia is freed to the soil, where it may be used by bacteria which oxidise it to nitrate (nitrification). Bacteria may on average add 25 pounds (11 kg) nitrogen per acre, and in an unfertilised field, this is the most important source of usable nitrogen. In a soil with 5% organic matter perhaps 2 to 5% of that is released to the soil by such decomposition. It occurs fastest in warm, moist, well aerated soil. The mineralisation of 3% of the organic material of a soil that is 4% organic matter overall, would release 120 pounds (54 kg) of nitrogen as ammonium per acre.

In nitrogen fixation, rhizobium bacteria convert N_2 to nitrate (NO_3^-). Rhizobia share a symbiotic relationship with host plants, since rhizobia supply the host with nitrogen and the host provides rhizobia with nutrients and a safe environment. It is estimated that such symbiotic bacteria in the root nodules of legumes add 45 to 250 pounds of nitrogen per acre per year, which may be sufficient for the crop. Other, free-living nitrogen-fixing bacteria and blue-green algae live independently in the soil and release nitrate when their dead bodies are converted by way of mineralisation.

Some amount of usable nitrogen is fixed by lightning as nitric oxide (NO) and nitrogen dioxide (NO_2^-). Nitrogen dioxide is soluble in water to form nitric acid (HNO_3) solution of H^+ and NO_3^-. Ammonia, NH_3, previously released from the soil or from combustion, may fall with precipitation as nitric acid at a rate of about five pounds nitrogen per acre per year.

Nitrogen Sequestration

When bacteria feed on soluble forms of nitrogen (ammonium and nitrate), they temporarily sequester that nitrogen in their bodies in a process called immobilisation. At a later time when those bacteria die, their nitrogen may be released as ammonium by the processes of mineralisation.

Protein material is easily broken down, but the rate of its decomposition is slowed by its attachment to the crystalline structure of clay and when trapped between the clay layers. The layers are small enough that bacteria cannot enter. Some organisms can exude extracellular enzymes that can act on the sequestered proteins. However, those enzymes too may be trapped on the clay crystals.

Ammonium fixation occurs when ammonium pushes potassium ions from between the layers of clay such as illite or montmorillonite. Only a small fraction of nitrogen is held this way.

Nitrogen Losses

Usable nitrogen may be lost from soils when it is in the form of nitrate, as it is easily leached. Further losses of nitrogen occur by denitrification, the process whereby soil bacteria convert nitrate (NO_3^-) to nitrogen gas, N_2 or N_2O. This occurs when poor soil aeration limits free oxygen, forcing bacteria to use the oxygen in nitrate for their respiratory process. Denitrification increases when oxidisable organic material is available and when soils are warm and slightly acidic. Denitrification may vary throughout a soil as the aeration varies from place to place. Denitrification may cause the loss of 10 to 20 percent of the available nitrates within a day and when conditions are favourable to that process, losses of up to 60 percent of nitrate applied as fertiliser may occur.

Ammonium volatilisation *occurs when ammonium reacts chemically with an alkaline soil, converting* NH_4^+ *to* NH_3. *The application of ammonium fertiliser to such a field can result in volatilisation losses of as much as 30 percent.*

Phosphorus

Phosphorus is the second most critical plant nutrient. The soil mineral apatite is the most common mineral source of phosphorus. While there is on average 1000 lb of phosphorus per acre in the soil, it is generally unavailable in the form of phosphates of low solubility. Total phosphorus is about 0.1 percent by weight of the soil, but only one percent of that is available. Of the part available, more than half comes from the mineralisation of organic matter. Agricultural fields may need to be fertilised to make up for the phosphorus that has been removed in the crop.

When phosphorus does form solubilised ions of $H_2PO_4^-$, they rapidly form insoluble phosphates of calcium or hydrous oxides of iron and aluminum. Phosphorus is largely immobile in the soil and is not leached but actually builds up in the surface layer if not cropped. The application of soluble fertilisers to soils may result in zinc deficiencies as zinc phosphates form. Conversely, the application of zinc to soils may immobilise phosphorus again as zinc phosphate. Lack of phosphorus may interfere with the normal opening of the plant leaf stomata, resulting in plant temperatures 10 percent higher than normal. Phosphorus is most available when soil pH is 6.5 in mineral soils and 5.5 in organic soils.

Potassium

The amount of potassium in a soil may be as much as 80,000 lb per acre-foot, of which only 150 lb is available for plant growth. Common mineral sources of potassium are the mica biotite and potassium feldspar, $KAlSi_3O_8$. When solubilised, half will be held as exchangeable cations on clay while the other half is in the soil water solution. Potassium fixation often occurs when soils dry and the potassium is bonded between layers of illite clay. Under certain conditions, dependent on the soil texture, intensity of drying, and initial amount of exchangeable potassium, the fixed percentage may be as much as 90 percent within ten minutes. Potassium may be leached from soils low in clay.

Calcium

Calcium is 1 percent by weight of soils and is generally available but may be low as it is soluble and can be leached. It is thus low in sandy and heavily leached soil or strongly acidic mineral soil. Calcium is supplied to the plant in the form of exchangeable ions and moderately soluble minerals. Calcium is more available on the soil colloids than is potassium because the common mineral calcite, $CaCO_3$, is more soluble than potassium-bearing minerals.

Magnesium

Magnesium is central to chlorophyll and aids in the uptake of phosphorus. The minimum amount of magnesium required for plant health is not sufficient for the health of forage animals. A common mineral source of magnesium is the black mica mineral, biotite. Magnesium is generally available in soil, but is missing from some along the Gulf and Atlantic coasts of the United States due to leaching by heavy precipitation.

Sulfur

Sulfur is essential to the formation of proteins and chlorophyll, and essential to plant vitamin synthesis. Most sulfur is made available to plants, like phosphorus, by its release from decomposing organic matter. Deficiencies may exist in some soils and if cropped, sulfur needs to be added. The application of large quantities of nitrogen to fields that have marginal amounts of sulfur may cause sulfur deficiency in the rapidly growing plants by the plant's growth outpacing the supply of sulfur. A 15-ton crop of onions uses up to 19 lb of sulfur and 4 tons of alfalfa uses 15 lb per acre. Sulfur abundance varies with depth. In a sample of soils in Ohio, United States, the sulfur abundance varied with depths, 0-6 inches, 6-12 inches, 12-18 inches, 18-24 inches in the amounts: 1056, 830, 686, 528 lb per acre respectively.

Micronutrients

The micronutrients essential for plant life, in their order of importance, include iron, manganese, zinc, copper, boron, chlorine and molybdenum. The term refers to plants' needs, not to their abundance in soil. They are required in very small amounts but are essential to plant health in that most are required parts of some enzyme system which speeds up plants' metabolisms. They are generally available in the mineral component of the soil, but the heavy application of phosphates can cause a deficiency in zinc and iron by the formation of insoluble zinc and iron phosphates. Iron deficiency may also result from excessive amounts of heavy metals or calcium minerals (lime) in the soil. Excess amounts of soluble boron, molybdenum and chloride are toxic.

Non-essential Nutrients

Nutrients which enhance the health but whose deficiency does not stop the life cycle of plants include: cobalt, strontium, vanadium, silicon and nickel. As their importance are evaluated they may be added to the list of essential plant nutrients.

Soil Organic Matter

Soil organic matter is made up of organic compounds and includes plant, animal and microbial material, both living and dead. A typical soil has a biomass composition of 70% microorganisms, 22% macrofauna, and 8% roots. The living component of an acre of soil may include 900 lb of earthworms, 2400 lb of fungi, 1500 lb of bacteria, 133 lb of protozoa and 890 lb of arthropods and algae.

A small part of the organic matter consists of the living cells such as bacteria, molds, and actinomycetes that work to break down the dead organic matter. Were it not for the action of these micro-organisms, the entire carbon dioxide part of the atmosphere would be sequestered as organic matter in the soil.

Chemically, organic matter is classed as follows:

- Polysaccharides
- cellulose
- hemicellulose
- starch

- pectin
- Lignins
- Proteins

Most living things in soils, including plants, insects, bacteria, and fungi, are dependent on organic matter for nutrients and energy. Soils have organic compounds in varying degrees of decomposition which rate is dependent on the temperature, soil moisture, and aeration. Bacteria and fungi feed on the raw organic matter, which are fed upon by amoebas, which in turn are fed upon by nematodes and arthropods. Organic matter holds soils open, allowing the infiltration of air and water, and may hold as much as twice its weight in water. Many soils, including desert and rocky-gravel soils, have little or no organic matter. Soils that are all organic matter, such as peat (histosols), are infertile. In its earliest stage of decomposition, the original organic material is often called raw organic matter. The final stage of decomposition is called humus.

In grassland, much of the organic matter added to the soil is from the deep, fibrous, grass root systems. By contrast, tree leaves falling on the forest floor are the principal source of soil organic matter in the forest. Another difference is the frequent occurrence in the grasslands of fires that destroy large amounts of aboveground material but stimulate even greater contributions from roots. Also, the much greater acidity under any forests inhibits the action of certain soil organisms that otherwise would mix much of the surface litter into the mineral soil. As a result, the soils under grasslands generally develop a thicker A horizon with a deeper distribution of organic matter than in comparable soils under forests, which characteristically store most of their organic matter in the forest floor (O horizon) and thin A horizon.

Humus

Humus refers to organic matter that has been decomposed by soil flora and fauna to the point where it is resistant to further breakdown. Humus usually constitutes only five percent of the soil or less by volume, but it is an essential source of nutrients and adds important textural qualities crucial to soil health and plant growth. Humus also hold bits of undecomposed organic matter which feed arthropods and worms which further improve the soil. The end product, humus, is soluble in water and forms a weak acid that can attack silicate minerals. Humus has a high cation exchange capacity that on a dry weight basis is many times greater than that of clay colloids. It also acts as a buffer, like clay, against changes in pH and soil moisture.

Humic acids and fulvic acids, which begin as raw organic matter, are important constituents of humus. After the death of plants and animals, microbes begin to feed on the residues, resulting finally in the formation of humus. With decomposition, there is a reduction of water-soluble constituents, cellulose and hemicellulose, and nutrients such as nitrogen, phosphorus, and sulfur. As the residues break down, only stable molecules made of aromatic carbon rings, oxygen and hydrogen remain in the form of humin, lignin and lignin complexes collectively called humus. While the structure of humus has few nutrients, it is able to attract and hold cation and anion nutrients by weak bonds that can be released into the soil solution in response to changes in soil pH.

Lignin is resistant to breakdown and accumulates within the soil. It also reacts with amino acids, which further increases its resistance to decomposition, including enzymatic decomposition by microbes. Fats and waxes from plant matter have some resistance to decomposition and persist in soils for a while. Clay soils often have higher organic contents that persist longer than soils without clay as the organic molecules adhere to and are stabilised by the clay. Proteins normally decompose readily, but when bound to clay particles, they become more resistant to decomposition. Clay particles also absorb the enzymes exuded by microbes which would normally break down proteins. The addition of organic matter to clay soils can render that organic matter and any added nutrients inaccessible to plants and microbes for many years. High soil tannin (polyphenol) content can cause nitrogen to be sequestered in proteins or cause nitrogen immobilisation.

Humus formation is a process dependent on the amount of plant material added each year and the type of base soil. Both are affected by climate and the type of organisms present. Soils with humus can vary in nitrogen content but typically have 3 to 6 percent nitrogen. Raw organic matter, as a reserve of nitrogen and phosphorus, is a vital component affecting soil fertility. Humus also absorbs water, and expands and shrinks between dry and wet states, increasing soil porosity. Humus is less stable than the soil's mineral constituents, as it is reduced by microbial decomposition, and over time its concentration diminishes without the addition of new organic matter. However, humus may persist over centuries if not millennia.

Climate and Organic Matter

The production, accumulation and degradation of organic matter are greatly dependent on climate. Temperature, soil moisture and topography are the major factors affecting the accumulation of organic matter in soils. Organic matter tends to accumulate under wet or cold conditions where decomposer activity is impeded by low temperature or excess moisture which results in anaerobic conditions.

Conversely, excessive rain and high temperatures of tropical climates enables rapid decomposition of organic matter and leaching of plant nutrients; forest ecosystems on these soils rely on efficient recycling of nutrients and plant matter to maintain their productivity. Excessive slope may encourage the erosion of the top layer of soil which holds most of the raw organic material that would otherwise eventually become humus.

Plant Residue in Soil

Cellulose and hemicellulose undergo fast decomposition by fungi and bacteria, with a half-life of 12–18 days in a temperate climate. Brown rot fungi can decompose the cellulose and hemicellulose, leaving the lignin and phenolic compounds behind. Starch, which is an energy storage system for plants, undergoes fast decomposition by bacteria and fungi. Lignin consists of polymers composed of 500 to 600 units with a highly branched, amorphous structure. Lignin undergoes very slow decomposition, mainly by white rot fungi and actinomycetes; its half-life under temperate conditions is about six months.

Soil Horizons

A horizontal layer of the soil, whose physical features, composition and age are distinct from those above and beneath, are referred to as a soil horizon. The naming of a horizon is based on the type of material of which it is composed. Those materials reflect the duration of specific processes of soil formation. They are labelled using a shorthand notation of letters and numbers which describe the horizon in terms of its colour, size, texture, structure, consistency, root quantity, pH, voids, boundary characteristics and presence of nodules or concretions. Few soil profiles have all the major horizons. Some may have only one horizon.

The exposure of parent material to favourable conditions produces mineral soils that are marginally suitable for plant growth. That growth often results in the accumulation of organic residues. The accumulated organic layer called the O horizon produces a more active soil due to the effect of the organisms that live within it. Organisms colonise and break down organic materials, making available nutrients upon which other plants and animals can live. After sufficient time, humus moves downward and is deposited in a distinctive organic surface layer called the A horizon.

Classification

Soil is classified into categories in order to understand relationships between different soils and to determine the suitability of a soil for a particular use. One of the first classification systems was developed by the Russian scientist Dokuchaev around 1880. It was modified a number of times by American and European researchers, and developed into the system commonly used until the 1960s. It was based on the idea that soils have a particular morphology based on the materials and factors that form them. In the 1960s, a different classification system began to emerge which focused on soil morphology instead of parental materials and soil-forming factors. Since then it has undergone further modifications. The World Reference Base for Soil Resources (WRB) aims to establish an international reference base for soil classification.

Soil Classification Systems

Australia

There are fourteen soil orders at the top level of the Australian Soil Classification. They are: Anthroposols, Organosols, Podosols, Vertosols, Hydrosols, Kurosols, Sodosols, Chromosols, Calcarosols, Ferrosols, Dermosols, Kandosols, Rudosols and Tenosols.

European Union

The EU's soil taxonomy is based on a new standard soil classification in the World Reference Base for Soil Resources produced by the UN's Food and Agriculture Organization. According to this, the major soils in the European Union are:

- Acrisols
- Albeluvisols
- Andosols
- Anthrosols
- Arenosols
- Calcisols
- Cambisols
- Chernozems
- Fluvisols
- Gleysols
- Gypsisols
- Histosols

- Kastanozems
- Leptosols
- Luvisols
- Phaeozems
- Planosols
- Podzols
- Regosols
- Solonchaks
- Solonetzes
- Umbrisols
- Vertisols

USA

A taxonomy is an arrangement in a systematic manner; the USDA soil taxonomy has six levels of classification. They are, from most general to specific: order, suborder, great group, subgroup, family and series. Soil properties that can be measured quantitatively are used in this classification system — they include: depth, moisture, temperature, texture, structure, cation exchange capacity, base saturation, clay mineralogy, organic matter content and salt content. There are 12 soil orders (the top hierarchical level) in the USDA soil taxonomy. The names of the orders end with the suffix *-sol*. The criteria for the different soil orders include properties that reflect major differences in the genesis of soils. The orders are:

- Alfisol - soils with aluminium and iron. They have horizons of clay accumulation, and form where there is enough moisture and warmth for at least three months of plant growth. They constitute 10.1% of soils worldwide.
- Andisols - volcanic ash soils. They are young and very fertile. They cover 1% of the world's ice-free surface.
- Aridisol - dry soils forming under desert conditions which have fewer than 90 consecutive days of moisture during the growing season and are nonleached. They include nearly 12% of soils on Earth. Soil formation is slow, and accumulated organic matter is scarce. They may have subsurface zones of caliche or duripan. Many aridisols have well-developed Bt horizons showing clay movement from past periods of greater moisture.
- Entisol - recently formed soils that lack well-developed horizons. Commonly found on unconsolidated river and beach sediments

of sand and clay or volcanic ash, some have an A horizon on top of bedrock. They are 18% of soils worldwide.

- Gelisols - permafrost soils with permafrost within two metres of the surface or gelic materials and permafrost within one metre. They constitute 9.1% of soils worldwide.
- Histosol - organic soils, formerly called bog soils, are 1.2% of soils worldwide.
- Inceptisol - young soils. They have subsurface horizon formation but show little eluviation and illuviation. They constitute 15% of soils worldwide.
- Mollisols - soft, deep, dark fertile soil formed in grasslands and some hardwood forests with very thick A horizons. They are 7% of soils worldwide.
- Oxisol - are heavily weathered, are rich in iron and aluminum oxides (sesquioxides) or kaolin but low in silica. They have only trace nutrients due to heavy tropical rainfall and high temperatures. They are 7.5% of soils worldwide.
- Spodosol - acid soils with organic colloid layer complexed with iron and aluminium leached from a layer above. They are typical soils of coniferous and deciduous forests in cooler climates. They constitute 4% of soils worldwide.
- Ultisol - acid soils in the humid tropics and subtropics, which are depleted in calcium, magnesium and potassium (important plant nutrients). They are highly weathered, but not as weathered as Oxisols. They make up 8% of the soil worldwide.
- Vertisol - inverted soils. They are clay-rich and tend to swell when wet and shrink upon drying, often forming deep cracks into which surface layers can fall. They are difficult to farm or to construct roads and buildings due to their high expansion rate. They constitute 2.4% of soils worldwide.
- The percentages listed above are for land area free of ice. "Soils of Mountains", which constitute the balance (11.6%), have a mixture of those listed above, or are classified as "Rugged Mountains" which have no soil.
- The above soil orders in sequence of increasing degree of development are Entisols, Inceptisols, Aridisols, Mollisols, Alfisols, Spodosols, Ultisols, and Oxisols. Histosols and Vertisols may appear in any of the above at any time during their development.

- The soil suborders within an order are differentiated on the basis of soil properties and horizons which depend on soil moisture and temperature. Forty-seven suborders are recognized in the United States.
- The soil great group category is a subdivision of a suborder in which the kind and sequence of soil horizons distinguish one soil from another. About 185 great groups are recognized in the United States. Horizons marked by clay, iron, humus and hard pans and soil features such as the expansion-contraction of clays (that produce self-mixing provided by clay), temperature, and marked quantities of various salts are used as distinguishing features.
- The great group categories are divided into three kinds of soil subgroups: typic, intergrade and extragrade. A typic subgroup represents the basic or 'typical' concept of the great group to which the described subgroup belongs. An intergrade subgroup describes the properties that suggest how it grades towards (is similar to) soils of other soil great groups, suborders or orders. These properties are not developed or expressed well enough to cause the soil to be included within the great group towards which they grade, but suggest similarities. Extragrade features are aberrant properties which prevent that soil from being included in another soil classification. About 1,000 soil subgroups are defined in the United States.
- A soil family category is a group of soils within a subgroup and describes the physical and chemical properties which affect the response of soil to agricultural management and engineering applications. The principal characteristics used to differentiate soil families include texture, mineralogy, pH, permeability, structure, consistency, the locale's precipitation pattern, and soil temperature. For some soils the criteria also specify the percentage of silt, sand and coarse fragments such as gravel, cobbles and rocks. About 4,500 soil families are recognised in the United States.
- A family may contain several soil series which describe the physical location using the name of a prominent physical feature such as a river or town near where the soil sample was taken. An example would be Merrimac for the Merrimack River in New Hampshire, USA. More than 14,000 soil series are recognised in the United States. This permits very specific descriptions of soils.

- A soil phase of series, originally called 'soil type' describes the soil surface texture, slope, stoniness, saltiness, erosion, and other conditions.

Uses

- Soil is used in agriculture, where it serves as the anchor and primary nutrient base for plants; however, as demonstrated by hydroponics, it is not essential to plant growth if the soil-contained nutrients can be dissolved in a solution. The types of soil and available moisture determine the species of plants that can be cultivated.
- Soil material is also a critical component in the mining, construction and landscape development industries. Soil serves as a foundation for most construction projects. The movement of massive volumes of soil can be involved in surface mining, road building and dam construction. Earth sheltering is the architectural practice of using soil for external thermal mass against building walls.
- Soil resources are critical to the environment, as well as to food and fibre production. Soil provides minerals and water to plants. Soil absorbs rainwater and releases it later, thus preventing floods and drought. Soil cleans water as it percolates through it. Soil is the habitat for many organisms: the major part of known and unknown biodiversity is in the soil, in the form of invertebrates (earthworms, woodlice, millipedes, centipedes, snails, slugs, mites, springtails, enchytraeids, nematodes, protists), bacteria, archaea, fungi and algae; and most organisms living above ground have part of them (plants) or spend part of their life cycle (insects) below-ground. Above-ground and below-ground biodiversities are tightly interconnected, making soil protection of paramount importance for any restoration or conservation plan.
- The biological component of soil is an extremely important carbon sink since about 57% of the biotic content is carbon. Even on desert crusts, cyanobacteria, lichens and mosses capture and sequester a significant amount of carbon by photosynthesis. Poor farming and grazing methods have degraded soils and released much of this sequestered carbon to the atmosphere. Restoring the world's soils could offset some of the huge increase in greenhouse gases causing global warming, while improving crop yields and reducing water needs.

- Waste management often has a soil component. Septic drain fields treat septic tank effluent using aerobic soil processes. Landfills use soil for daily cover. Land application of waste water relies on soil biology to aerobically treat BOD.
- Organic soils, especially peat, serve as a significant fuel resource; but wide areas of peat production, such as sphagnum bogs, are now protected because of patrimonial interest.
- Geophagy is the practice of eating soil-like substances. Both animals and human cultures occasionally consume soil for medicinal, recreational, or religious purposes. It has been shown that some monkeys consume soil, together with their preferred food (tree foliage and fruits), in order to alleviate tannin toxicity.
- Soils filter and purify water and affect its chemistry. Rain water and pooled water from ponds, lakes and rivers percolate through the soil horizons and the upper rock strata, thus becoming groundwater. Pests (viruses) and pollutants, such as persistent organic pollutants (chlorinated pesticides, polychlorinated biphenyls), oils (hydrocarbons), heavy metals (lead, zinc, cadmium), and excess nutrients (nitrates, sulfates, phosphates) are filtered out by the soil. Soil organisms metabolise them or immobilise them in their biomass and necromass, thereby incorporating them into stable humus. The physical integrity of soil is also a prerequisite for avoiding landslides in rugged landscapes.

Degradation

- Land degradation refers to a human-induced or natural process which impairs the capacity of land to function. Soils are the critical component in land degradation when it involves acidification, contamination, desertification, erosion or salination.
- While soil acidification is beneficial in the case of alkaline soils, it degrades land when it lowers crop productivity and increases soil vulnerability to contamination and erosion. Soils are often initially acid because their parent materials were acid and initially low in the basic cations (calcium, magnesium, potassium and sodium). Acidification occurs when these elements are leached from the soil profile by rainfall or the by harvesting of forest or agricultural crops. Soil acidification is accelerated by the use of acid-forming nitrogenous fertilizers and by the effects of acid precipitation.

- Soil contamination at low levels is often within soil's capacity to treat and assimilate waste material. Soil biota can treat waste by transforming it; soil colloids can adsorb waste material. Many waste treatment processes rely on this treatment capacity. Exceeding treatment capacity can damage soil biota and limit soil function. Derelict soils occur where industrial contamination or other development activity damages the soil to such a degree that the land cannot be used safely or productively. Remediation of derelict soil uses principles of geology, physics, chemistry and biology to degrade, attenuate, isolate or remove soil contaminants to restore soil functions and values. Techniques include leaching, air sparging, chemical amendments, phytoremediation, bioremediation and natural degradation.
- Desertification is an environmental process of ecosystem degradation in arid and semi-arid regions, often caused by human activity. It is a common misconception that droughts cause desertification. Droughts are common in arid and semiarid lands. Well-managed lands can recover from drought when the rains return. Soil management tools include maintaining soil nutrient and organic matter levels, reduced tillage and increased cover. These practices help to control erosion and maintain productivity during periods when moisture is available. Continued land abuse during droughts, however, increases land degradation. Increased population and livestock pressure on marginal lands accelerates desertification.
- Erosion of soil is caused by water, wind, ice, and movement in response to gravity. More than one kind of erosion can occur simultaneously. Erosion is distinguished from weathering, since erosion also transports eroded soil away from its place of origin (soil in transit may be described as sediment). Erosion is an intrinsic natural process, but in many places it is greatly increased by human activity, especially poor land use practices. These include agricultural activities which leave the soil bare during times of heavy rain or strong winds, overgrazing, deforestation, and improper construction activity. Improved management can limit erosion. Soil conservation techniques which are employed include changes of land use (such as replacing erosion-prone crops with grass or other soil-binding plants), changes to the timing or type of agricultural operations, terrace building, use of erosion-suppressing cover materials (including cover crops and other plants), limiting disturbance during construction, and avoiding construction during erosion-prone periods.

- A serious and long-running water erosion problem occurs in China, on the middle reaches of the Yellow River and the upper reaches of the Yangtze River. From the Yellow River, over 1.6 billion tons of sediment flow each year into the ocean. The sediment originates primarily from water erosion (gully erosion) in the Loess Plateau region of northwest China.
- Soil piping is a particular form of soil erosion that occurs below the soil surface. It causes levee and dam failure, as well as sink hole formation. Turbulent flow removes soil starting at the mouth of the seep flow and the subsoil erosion advances up-gradient. The term sand boil is used to describe the appearance of the discharging end of an active soil pipe.
- Soil salination is the accumulation of free salts to such an extent that it leads to degradation of the agricultural value of soils and vegetation. Consequences include corrosion damage, reduced plant growth, erosion due to loss of plant cover and soil structure, and water quality problems due to sedimentation. Salination occurs due to a combination of natural and human-caused processes. Arid conditions favour salt accumulation. This is especially apparent when soil parent material is saline. Irrigation of arid lands is especially problematic. All irrigation water has some level of salinity. Irrigation, especially when it involves leakage from canals and overirrigation in the field, often raises the underlying water table. Rapid salination occurs when the land surface is within the capillary fringe of saline groundwater. Soil salinity control involves watertable control and flushing with higher levels of applied water in combination with tile drainage or another form of subsurface drainage.

Reclamation

Soils which contain high levels of particular clays, such as smectites, are often very fertile. For example, the smectite-rich clays of Thailand's Central Plains are among the most productive in the world.

Many farmers in tropical areas, however, struggle to retain organic matter in the soils they work. In recent years, for example, productivity has declined in the low-clay soils of northern Thailand. Farmers initially responded by adding organic matter from termite mounds, but this was unsustainable in the long-term. Scientists experimented with adding bentonite, one of the smectite family of clays, to the soil. In field trials, conducted by scientists from the International Water

Management Institute in cooperation with Khon Kaen University and local farmers, this had the effect of helping retain water and nutrients. Supplementing the farmer's usual practice with a single application of 200 kg bentonite per rai (6.26 rai = 1 hectare) resulted in an average yield increase of 73%. More work showed that applying bentonite to degraded sandy soils reduced the risk of crop failure during drought years.

In 2008, three years after the initial trials, IWMI scientists conducted a survey among 250 farmers in northeast Thailand, half of whom had applied bentonite to their fields. The average improvement for those using the clay addition was 18% higher than for non-clay users. Using the clay had enabled some farmers to switch to growing vegetables, which need more fertile soil. This helped to increase their income. The researchers estimated that 200 farmers in northeast Thailand and 400 in Cambodia had adopted the use of clays, and that a further 20,000 farmers were introduced to the new technique.

If the soil is too high in clay, adding gypsum, washed river sand and organic matter will balance the composition. Adding organic matter (like ramial chipped wood for instance) to soil which is depleted in nutrients and too high in sand will boost its quality.

Chapter 3

Water Management

Water is a transparent fluid which forms the world's streams, lakes, oceans and rain, and is the major constituent of the fluids of living things. As a chemical compound, a water molecule contains one oxygen and two hydrogen atoms that are connected by covalent bonds. Water is a liquid at standard ambient temperature and pressure, but it often co-exists on Earth with its solid state, ice; and gaseous state, steam (water vapor).

Water covers 71% of the Earth's surface. It is vital for all known forms of life. On Earth, 96.5% of the planet's water is found in seas and oceans, 1.7% in groundwater, 1.7% in glaciers and the ice caps of Antarctica and Greenland, a small fraction in other large water bodies, and 0.001% in the air as vapor, clouds (formed of solid and liquid water particles suspended in air), and precipitation. Only 2.5% of the Earth's water is freshwater, and 98.8% of that water is in ice and groundwater. Less than 0.3% of all freshwater is in rivers, lakes, and the atmosphere, and an even smaller amount of the Earth's freshwater (0.003%) is contained within biological bodies and manufactured products.

Water on Earth moves continually through the water cycle of evaporation and transpiration (evapotranspiration), condensation, precipitation, and runoff, usually reaching the sea. Evaporation and transpiration contribute to the precipitation over land. Water used in the production of a good or service is known as virtual water.

Safe drinking water is essential to humans and other lifeforms even though it provides no calories or organic nutrients. Access to safe drinking water has improved over the last decades in almost every part of the world, but approximately one billion people still lack access to safe water and over 2.5 billion lack access to adequate sanitation. There is a clear correlation between access to safe water and gross domestic product per capita. However, some observers have estimated that by 2025 more than half of the world population will be facing water-based vulnerability. A report, issued in November 2009, suggests that by 2030, in some developing regions of the world, water demand will exceed supply by 50%. Water plays an important role in the world economy, as it functions as a solvent for a wide variety of chemical substances and facilitates industrial cooling and transportation. Approximately 70% of the fresh water used by humans goes to agriculture.

Chemical and Physical Properties

Water is the chemical substance with chemical formula H_2O: one molecule of water has two hydrogen atoms covalently bonded to a single oxygen atom. Water appears in nature in all three common states of matter (solid, liquid, and gas) and may take many different forms on Earth: water vapor and clouds in the sky, seawater in the oceans, icebergs in the polar oceans, glaciers in the mountains, fresh and salt water lakes, rivers, and aquifers in the ground.

The major chemical and physical properties of water are:

- Water is a liquid at standard temperature and pressure. It is tasteless and odorless. The intrinsic colour of water and ice is a very slight blue hue, although both appear colourless in small quantities. Water vapour is essentially invisible as a gas.
- Water is transparent in the visible electromagnetic spectrum. Thus aquatic plants can live in water because sunlight can reach them. Infrared light is strongly absorbed by the hydrogen-oxygen or OH bonds.
- Since the water molecule is not linear and the oxygen atom has a higher electronegativity than hydrogen atoms, the oxygen atom carries a slight negative charge, whereas the hydrogen atoms are slightly positive. As a result, water is a polar molecule with an electrical dipole moment. Water also can form an unusually large number of intermolecular hydrogen bonds (four) for a molecule of its size. These factors lead to strong attractive forces between molecules of water, giving rise to water's high surface tension and capillary forces. The capillary action refers to the tendency of water to move up a narrow tube against the force of gravity. This property is relied upon by all vascular plants, such as trees.

- Water is a good polar solvent and is often referred to as *the universal solvent*. Substances that dissolve in water, e.g., salts, sugars, acids, alkalis, and some gases – especially oxygen and carbon dioxide (carbonation) – are known as *hydrophilic* (water-loving) substances, while those that are immiscible with water (e.g., fats and oils), are known as *hydrophobic* (water-fearing) substances.
- All of the components in cells (proteins, DNA and polysaccharides) are dissolved in water, deriving their structure and activity from their interactions with the water.
- Pure water has a low electrical conductivity, but this increases with the dissolution of a small amount of ionic material such as sodium chloride.
- The boiling point of water (and all other liquids) is dependent on the barometric pressure. For example, on the top of Mount Everest water boils at 68 °C (154 °F), compared to 100 °C (212 °F) at sea level at a similar latitude (since latitude modifies atmospheric pressure slightly). Conversely, water deep in the ocean near geothermal vents can reach temperatures of hundreds of degrees and remain liquid.
- At 4181.3 J/(kg·K), water has a high specific heat capacity, as well as a high heat of vaporization (40.65 kJ mol^{-1}), both of which are a result of the extensive hydrogen bonding between its molecules. These two unusual properties allow water to moderate Earth's climate by buffering large fluctuations in temperature.
- The density of liquid water is 1,000 kg/m^3 (62.43 lb/cu ft) at 4 °C. Ice has a density of 917 kg/m^3 (57.25 lb/cu ft).

Figure: *ADR label for transporting goods dangerously reactive with water*

- The maximum density of water occurs at 3.98 °C (39.16 °F). Most known pure substances become more dense as they cool, however water has the anomalous property of becoming less dense when it is cooled to its solid form, ice.

  During cooling water becomes more dense until reaching 3.98 °C. Below this temperature, the open structure of ice is gradually formed in the low temperature water; the random orientations of the water molecules in the liquid are maintained by the thermal motion, and below 3.98 °C there is not enough thermal energy to maintain this randomness.

  As water is cooled there are two competing effects:

  1) decreasing volume, and
  2) increase overall volume of the liquid as the molecules begin to orient into the organized structure of ice. Between 3.98 °C and 0 °C, the second effect will cancel the first effect so the net effect is an increase of volume with decreasing temperature.

  Water expands to occupy a 9% greater volume as ice, which accounts for the fact that ice floats on liquid water, as in icebergs.

- Water is miscible with many liquids, such as ethanol, in all proportions, forming a single homogeneous liquid. On the other hand, water and most oils are immiscible, usually forming layers with the least dense liquid as the top layer, and the most dense layer at the bottom.
- Water forms an azeotrope with many other solvents.
- Water can be split by electrolysis into hydrogen and oxygen. The energy required to split water into hydrogen and oxygen by electrolysis or any other means is greater than the energy that can be collected when the hydrogen and oxygen recombine.
- As an oxide of hydrogen, water is formed when hydrogen or hydrogen-containing compounds burn or react with oxygen or oxygen-containing compounds. Water is not a fuel, it is an end-product of the combustion of hydrogen.
- Elements which are more electropositive than hydrogen such as lithium, sodium, calcium, potassium and caesium displace hydrogen from water, forming hydroxides. Being a flammable gas, the hydrogen given off is dangerous and the reaction of water with the more electropositive of these elements may be violently explosive.

Property	Remarks	Importance to the environment
Physical state	Only substance occurring naturally in all three phases as solid, liquid, and gas on Earth's surface	Transfer of heat between ocean and atmosphere by phase change
Dissolving ability	Dissolves more substances in greater quantities than any other common liquid	Important in chemical, physical, and biological processes
Density: mass per unit volume	Density is determined by (1) temperature, (2) salinity, and (3) pressure, in that order of importance. The temperature of maximum density for pure water is 4 °C. For seawater, the freezing point decreases with increasing salinity	Controls oceanic vertical circulation, aids in heat distribution, and allows seasonal stratification
Surface tension	Highest of all common liquids	Controls drop formation in rain and clouds; important in cell physiology
Conduction of heat	Highest of all common liquids	Important on the small scale, especially on cellular level
Heat capacity	Highest of all common solids and liquids	Prevents extreme range in Earth's temperatures (i.e., great heat moderator)
Latent heat of fusion	Highest of all common liquids and most solids	Thermostatic heat-regulating effect due to the release of heat on freezing and absorption on melting
Latent heat of vaporization	Highest of all common substances	Immense importance: a major factor in the transfer of heat in and between ocean and atmosphere, driving weather and climate
Refractive index	Increases with increasing salinity and decreases with increasing temperature	Objects appear closer than in air
Transparency	Relatively great for visible light; absorption high for infrared and ultraviolet	Important for photosynthesis
Sound transmission	Good compared with other fluids	Allows for sonar and precision depth recorders to rapidly determine water depth, and to detect subsurface features and animals; sounds can be heard great distances underwater
Compressibility	Only slight	Density changes only slightly with pressure/depth
Boiling and melting points	Unusually high	Allows water to exist as a liquid on most of Earth

Taste and odor

Pure H_2O is tasteless and odorless. Water can dissolve many different substances, giving it varying tastes and odors. Humans, and other animals, have developed senses that enable them to evaluate the potability of water by avoiding water that is too salty or putrid.

The taste of spring water and mineral water, often advertised in marketing of consumer products, derives from the minerals dissolved in it. The advertised purity of spring and mineral water refers to absence of toxins, pollutants, and microbes, not to the absence of naturally occurring minerals.

Distribution in Nature

In the Universe

Much of the universe's water is produced as a byproduct of star formation. When stars are born, their birth is accompanied by a strong outward wind of gas and dust. When this outflow of material eventually impacts the surrounding gas, the shock waves that are created compress and heat the gas. The water observed is quickly produced in this warm dense gas.

On 22 July 2011 a report described the discovery of a gigantic cloud of water vapor containing "140 trillion times more water than all of Earth's oceans combined" around a quasar located 12 billion light years from Earth. According to the researchers, the "discovery shows that water has been prevalent in the universe for nearly its entire existence".

Water has been detected in interstellar clouds within our galaxy, the Milky Way. Water probably exists in abundance in other galaxies, too, because its components, hydrogen and oxygen, are among the most abundant elements in the universe. Based on models of the formation and evolution of the Solar System and that of other star systems, most other planetary systems are likely to have similar ingredients.

Water Vapor

Water is present as vapor in:

- Atmosphere of the Sun: in detectable trace amounts
- Atmosphere of Mercury: 3.4%, and large amounts of water in Mercury's exosphere
- Atmosphere of Venus: 0.002%
- Earth's atmosphere: ~0.40% over full atmosphere, typically 1–4% at surface; as well as that of the Moon in trace amounts
- Atmosphere of Mars: 0.03%
- Atmosphere of Ceres
- Atmosphere of Jupiter: 0.0004% – in ices only; and that of its moon Europa
- Atmosphere of Saturn – in ices only; and that of its moons Titan (stratospheric), Enceladus: 91% and Dione (exosphere)
- Atmosphere of Uranus - in trace amounts below 50 bar
- Atmosphere of Neptune - found in the deeper layers
- Extrasolar planet atmospheres: including those of HD 189733 b and HD 209458 b, Tau Boötis b and HAT-P-11b
- Circumstellar disks: including those of TW Hydrae, IRC +10216 and APM 08279+5255

Liquid Water

Liquid water is known to be present on Earth, covering 71% of our planet's surface. Scientists believe liquid water is present in the Saturnian moons of Enceladus, as a 10 kilometre thick ocean approximately 30-40 kilometres below Enceladus' south polar surface, and Titan, as a subsurface layer, possibly mixed with ammonia. Liquid water may also exist on Jupiter's moon Ganymede as a layer sandwiched between high pressure ice and rock.

Water Ice

Water is present as ice on:

- Mars: under the regolith and at the poles
- Earth-Moon system: mainly as ice sheets on Earth and in Lunar craters and volcanic rocks NASA reported the detection of water molecules by NASA's Moon Mineralogy Mapper aboard the Indian Space Research Organization's Chandrayaan-1 spacecraft in September 2009.
- Jupiter's moons: Europa's surface and also that of Ganymede
- Saturn: in the planets ring system and on the surface and mantle of Titan and Enceladus
- Pluto-Charon system
- Comets and related (Kuiper belt and Oort cloud objects).

And may also be present on:

- Mercury's poles
- Ceres
- Tethys

Exotic Forms

Water and other volatiles probably comprise much of the internal structures of Uranus and Neptune and the water in the deeper layers may be in the form of ionic water in which the molecules break down into a soup of hydrogen and oxygen ions, and deeper down as superionic water in which the oxygen crystallises but the hydrogen ions float around freely within the oxygen lattice.

Water and Habitable Zone

The existence of liquid water, and to a lesser extent its gaseous and solid forms, on Earth are vital to the existence of life on Earth as we know it. The Earth is located in the habitable zone of the solar system; if it were slightly closer to or farther from the Sun (about 5%, or about 8 million kilometres), the conditions which allow the three forms to be present simultaneously would be far less likely to exist.

Earth's gravity allows it to hold an atmosphere. Water vapor and carbon dioxide in the atmosphere provide a temperature buffer (greenhouse effect) which helps maintain a relatively steady surface temperature. If Earth were smaller, a thinner atmosphere would allow temperature extremes, thus preventing the accumulation of water except in polar ice caps (as on Mars).

The surface temperature of Earth has been relatively constant through geologic time despite varying levels of incoming solar radiation (insolation), indicating that a dynamic process governs Earth's temperature via a combination of greenhouse gases and surface or atmospheric albedo. This proposal is known as the *Gaia hypothesis*.

The state of water on a planet depends on ambient pressure, which is determined by the planet's gravity. If a planet is sufficiently massive, the water on it may be solid even at high temperatures, because of the high pressure caused by gravity, as it was observed on exoplanets Gliese 436 b and GJ 1214 b.

On Earth

Hydrology is the study of the movement, distribution, and quality of water throughout the Earth. The study of the distribution of water is hydrography. The study of the distribution and movement of groundwater is hydrogeology, of glaciers is glaciology, of inland waters is limnology and distribution of oceans is oceanography. Ecological processes with hydrology are in focus of ecohydrology. The collective mass of water found on, under, and over the surface of a planet is called the hydrosphere. Earth's approximate water volume (the total water supply of the world) is 1,338,000,000 km^3 (321,000,000 mi^3).

Liquid water is found in bodies of water, such as an ocean, sea, lake, river, stream, canal, pond, or puddle. The majority of water on Earth is sea water. Water is also present in the atmosphere in solid, liquid, and vapor states. It also exists as groundwater in aquifers.

Water is important in many geological processes. Groundwater is present in most rocks, and the pressure of this groundwater affects patterns of faulting. Water in the mantle is responsible for the melt that produces volcanoes at subduction zones. On the surface of the Earth, water is important in both chemical and physical weathering processes. Water, and to a lesser but still significant extent, ice, are also responsible for a large amount of sediment transport that occurs on the surface of the earth. Deposition of transported sediment forms many types of sedimentary rocks, which make up the geologic record of Earth history.

Water Cycle

The water cycle (known scientifically as the hydrologic cycle) refers to the continuous exchange of water within the hydrosphere, between the atmosphere, soil water, surface water, groundwater, and plants.

Water moves perpetually through each of these regions in the *water cycle* consisting of following transfer processes:

- evaporation from oceans and other water bodies into the air and transpiration from land plants and animals into air.
- precipitation, from water vapor condensing from the air and falling to earth or ocean.
- runoff from the land usually reaching the sea.

Most water vapor over the oceans returns to the oceans, but winds carry water vapor over land at the same rate as runoff into the sea, about 47 Tt per year. Over land, evaporation and transpiration contribute another 72 Tt per year. Precipitation, at a rate of 119 Tt per year over land, has several forms: most commonly rain, snow, and hail, with some contribution from fog and dew. Dew is small drops of water that are condensed when a high density of water vapor meets a cool surface. Dew usually forms in the morning when the temperature is the lowest, just before sunrise and when the temperature of the earth's surface starts to increase. Condensed water in the air may also refract sunlight to produce rainbows.

Water runoff often collects over watersheds flowing into rivers. A mathematical model used to simulate river or stream flow and calculate water quality parameters is a hydrological transport model. Some water is diverted to irrigation for agriculture. Rivers and seas offer opportunity for travel and commerce. Through erosion, runoff shapes the environment creating river valleys and deltas which provide rich soil and level ground for the establishment of population centres. A flood occurs when an area of land, usually low-lying, is covered with water. It is when a river overflows its banks or flood comes from the sea. A drought is an extended period of months or years when a region notes a deficiency in its water supply. This occurs when a region receives consistently below average precipitation.

Fresh Water Storage

Some runoff water is trapped for periods of time, for example in lakes. At high altitude, during winter, and in the far north and south, snow collects in ice caps, snow pack and glaciers. Water also infiltrates the ground and goes into aquifers. This groundwater later flows back to the surface in springs, or more spectacularly in hot springs and geysers. Groundwater is also extracted artificially in wells. This water storage is important, since clean, fresh water is essential to human and other land-based life. In many parts of the world, it is in short supply.

Sea Water

Sea water contains about 3.5% salt on average, plus smaller amounts of other substances. The physical properties of sea water differ from fresh water in some important respects. It freezes at a lower temperature (about –1.9 °C) and its density increases with decreasing temperature to the freezing point, instead of reaching maximum density at a temperature above freezing. The salinity of water in major seas varies from about 0.7% in the Baltic Sea to 4.0% in the Red Sea.

Tides

Tides are the cyclic rising and falling of local sea levels caused by the tidal forces of the Moon and the Sun acting on the oceans. Tides cause changes in the depth of the marine and estuarine water bodies and produce oscillating currents known as tidal streams. The changing tide produced at a given location is the result of the changing positions of the Moon and Sun relative to the Earth coupled with the effects of Earth rotation and the local bathymetry. The strip of seashore that is submerged at high tide and exposed at low tide, the intertidal zone, is an important ecological product of ocean tides.

Effects on Life

From a biological standpoint, water has many distinct properties that are critical for the proliferation of life that set it apart from other substances. It carries out this role by allowing organic compounds to react in ways that ultimately allow replication. All known forms of life depend on water. Water is vital both as a solvent in which many of the body's solutes dissolve and as an essential part of many metabolic processes within the body. Metabolism is the sum total of anabolism and catabolism. In anabolism, water is removed from molecules (through energy requiring enzymatic chemical reactions) in order to grow larger molecules (e.g. starches, triglycerides and proteins for storage of fuels and information). In catabolism, water is used to break bonds in order to generate smaller molecules (e.g. glucose, fatty acids and amino acids to be used for fuels for energy use or other purposes). Without water, these particular metabolic processes could not exist.

Water is fundamental to photosynthesis and respiration. Photosynthetic cells use the sun's energy to split off water's hydrogen from oxygen. Hydrogen is combined with CO_2 (absorbed from air or water) to form glucose and release oxygen. All living cells use such fuels and oxidize the hydrogen and carbon to capture the sun's energy and reform water and CO_2 in the process (cellular respiration).

Water is also central to acid-base neutrality and enzyme function. An acid, a hydrogen ion (H^+, that is, a proton) donor, can be neutralized by a base, a proton acceptor such as a hydroxide ion (OH^e) to form water. Water is considered to be neutral, with a pH (the negative log of the hydrogen ion concentration) of 7. Acids have pH values less than 7 while bases have values greater than 7.

Aquatic Life Forms

Earth surface waters are filled with life. The earliest life forms appeared in water; nearly all fish live exclusively in water, and there are many types of marine mammals, such as dolphins and whales. Some kinds of animals, such as amphibians, spend portions of their lives in water and portions on land. Plants such as kelp and algae grow in the water and are the basis for some underwater ecosystems. Plankton is generally the foundation of the ocean food chain.

Aquatic vertebrates must obtain oxygen to survive, and they do so in various ways. Fish have gills instead of lungs, although some species of fish, such as the lungfish, have both. Marine mammals, such as dolphins, whales, otters, and seals need to surface periodically to breathe air. Some amphibians are able to absorb oxygen through their skin. Invertebrates exhibit a wide range of modifications to survive in poorly oxygenated waters including breathing tubes and gills (*Carcinus*). However as invertebrate life evolved in an aquatic habitat most have little or no specialisation for respiration in water.

Effects on Human Civilization

Figure: *Water fountain*

Civilization has historically flourished around rivers and major waterways; Mesopotamia, the so-called cradle of civilization, was situated between the major rivers Tigris and Euphrates; the ancient society of the Egyptians depended entirely upon the Nile. Large metropolises like Rotterdam, London, Montreal, Paris, New York City, Buenos Aires, Shanghai, Tokyo, Chicago, and Hong Kong owe their success in part to their easy accessibility via water and the resultant expansion of trade. Islands with safe water ports, like Singapore, have flourished for the same reason. In places such as North Africa and the Middle East, where water is more scarce, access to clean drinking water was and is a major factor in human development.

Health and Pollution

Water fit for human consumption is called drinking water or potable water. Water that is not potable may be made potable by filtration or distillation, or by a range of other methods.

Water that is not fit for drinking but is not harmful for humans when used for swimming or bathing is called by various names other than potable or drinking water, and is sometimes called safe water, or "safe for bathing". Chlorine is a skin and mucous membrane irritant that is used to make water safe for bathing or drinking. Its use is highly technical and is usually monitored by government regulations (typically 1 part per million (ppm) for drinking water, and 1–2 ppm of chlorine not yet reacted with impurities for bathing water). Water for bathing may be maintained in satisfactory microbiological condition using chemical disinfectants such as chlorine or ozone or by the use of ultraviolet light.

In the USA, non-potable forms of wastewater generated by humans may be referred to as greywater, which is treatable and thus easily able to be made potable again, and blackwater, which generally contains sewage and other forms of waste which require further treatment in order to be made reusable. Greywater composes 50–80% of residential wastewater generated by a household's sanitation equipment (sinks, showers and kitchen runoff, but not toilets, which generate blackwater.) These terms may have different meanings in other countries and cultures.

This natural resource is becoming scarcer in certain places, and its availability is a major social and economic concern. Currently, about a billion people around the world routinely drink unhealthy water. Most countries accepted the goal of halving by 2015 the number of people worldwide who do not have access to safe water and sanitation during the 2003 G8 Evian summit. Even if this difficult goal is met, it will still leave more than an estimated half a billion people without access to safe drinking water and over a billion without access to adequate sanitation. Poor water quality and bad sanitation are deadly; some five million deaths a year are caused by polluted drinking water.

The World Health Organization estimates that safe water could prevent 1.4 million child deaths from diarrhea each year.

Water, however, is not a finite resource, but rather re-circulated as potable water in precipitation in quantities many degrees of magnitude higher than human consumption. Therefore, it is the relatively small quantity of water in reserve in the earth (about 1% of our drinking water supply, which is replenished in aquifers around every 1 to 10 years), that is a non-renewable resource, and it is, rather, the distribution of potable and irrigation water which is scarce, rather than the actual amount of it that exists on the earth. Water-poor countries use importation of goods as the primary method of importing water (to leave enough for local human consumption), since the manufacturing process uses around 10 to 100 times products' masses in water.

In the developing world, 90% of all wastewater still goes untreated into local rivers and streams. Some 50 countries, with roughly a third of the world's population, also suffer from medium or high water stress, and 17 of these extract more water annually than is recharged through their natural water cycles. The strain not only affects surface freshwater bodies like rivers and lakes, but it also degrades groundwater resources.

Human Uses

Agriculture

The most important use of water in agriculture is for irrigation, which is a key component to produce enough food. Irrigation takes up to 90% of water withdrawn in some developing countries and significant proportions in more economically developed countries (United States, 30% of freshwater usage is for irrigation).

Fifty years ago, the common perception was that water was an infinite resource. At this time, there were fewer than half the current number of people on the planet. People were not as wealthy as today, consumed fewer calories and ate less meat, so less water was needed to produce their food. They required a third of the volume of water we presently take from rivers. Today, the competition for the fixed amount of water resources is much more intense, giving rise to the concept of peak water. This is because there are now nearly seven billion people on the planet, their consumption of water-thirsty meat and vegetables is rising, and there is increasing competition for water from industry, urbanisation and biofuel crops. In future, even more water will be needed to produce food because the Earth's population is forecast to rise to 9 billion by 2050.

An assessment of water management in agriculture was conducted in 2007 by the International Water Management Institute in Sri Lanka to see if the world had sufficient water to provide food for its growing population. It assessed the current availability of water for agriculture on a global scale and mapped out locations suffering from water scarcity. It found that a fifth of the world's people, more than 1.2 billion, live in areas of physical water scarcity, where there is not enough water to meet all demands. A further 1.6 billion people live in areas experiencing economic water scarcity, where the lack of investment in water or insufficient human capacity make it impossible for authorities to satisfy the demand for water. The report found that it would be possible to produce the food required in future, but that continuation of today's food production and environmental trends would lead to crises in many parts of the world. To avoid a global water crisis, farmers will have to strive to increase productivity to meet growing demands for food, while industry and cities find ways to use water more efficiently.

As a Scientific Standard

On 7 April 1795, the gram was defined in France to be equal to "the absolute weight of a volume of pure water equal to a cube of one hundredth of a metre, and at the temperature of melting ice." For practical purposes though, a metallic reference standard was required, one thousand times more massive, the kilogram. Work was therefore commissioned to determine precisely the mass of one liter of water. In spite of the fact that the decreed definition of the gram specified water at 0 °C — a highly reproducible *temperature* — the scientists chose to redefine the standard and to perform their measurements at the temperature of highest water *density*, which was measured at the time as 4 °C (39 °F).

The Kelvin temperature scale of the SI system is based on the triple point of water, defined as exactly 273.16 K or 0.01 °C. The scale is an absolute temperature scale with the same increment as the Celsius temperature scale, which was originally defined according the boiling point (set to 100 °C) and melting point (set to 0 °C) of water.

Natural water consists mainly of the isotopes hydrogen-1 and oxygen-16, but there is also a small quantity of heavier isotopes such as hydrogen-2 (deuterium). The amount of deuterium oxides or heavy water is very small, but it still affects the properties of water. Water from rivers and lakes tends to contain less deuterium than seawater. Therefore, standard water is defined in the Vienna Standard Mean Ocean Water specification.

For Drinking

The human body contains from 55% to 78% water, depending on body size. To function properly, the body requires between one and seven liters of water per day to avoid dehydration; the precise amount depends on the level of activity, temperature, humidity, and other factors. Most of this is ingested through foods or beverages other than drinking straight water. It is not clear how much water intake is needed by healthy people, though most specialists agree that approximately 2 liters (6 to 7 glasses) of water daily is the minimum to maintain proper hydration. Medical literature favours a lower consumption, typically 1 liter of water for an average male, excluding extra requirements due to fluid loss from exercise or warm weather.

For those who have healthy kidneys, it is rather difficult to drink too much water, but (especially in warm humid weather and while exercising) it is dangerous to drink too little. People can drink far more water than necessary while exercising, however, putting them at risk of water intoxication (hyperhydration), which can be fatal. The popular claim that "a person should consume eight glasses of water per day" seems to have no real basis in science. Similar misconceptions concerning the effect of water on weight loss and constipation have also been dispelled.

Figure: *Hazard symbol for non-potable water*

An original recommendation for water intake in 1945 by the Food and Nutrition Board of the United States National Research Council read: "An ordinary standard for diverse persons is 1 milliliter for each calorie of food. Most of this quantity is contained in prepared foods." The latest dietary reference intake report by the United States National Research Council in general recommended (including food sources): 3.7 liters for men and 2.7 liters of water total for women.

Specifically, pregnant and breastfeeding women need additional fluids to stay hydrated. The Institute of Medicine (U.S.) recommends that, on average, men consume 3.0 liters and women 2.2 liters; pregnant women should increase intake to 2.4 liters (10 cups) and breastfeeding women should get 3 liters (12 cups), since an especially large amount of fluid is lost during nursing. Also noted is that normally, about 20% of water intake comes from food, while the rest comes from drinking water and beverages (caffeinated included). Water is excreted from the body in multiple forms; through urine and feces, through sweating, and by exhalation of water vapor in the breath. With physical exertion and heat exposure, water loss will increase and daily fluid needs may increase as well.

Humans require water with few impurities. Common impurities include metal salts and oxides, including copper, iron, calcium and lead, and/or harmful bacteria, such as *Vibrio*. Some solutes are acceptable and even desirable for taste enhancement and to provide needed electrolytes. The single largest (by volume) freshwater resource suitable for drinking is Lake Baikal in Siberia.

Washing

The propensity of water to form solutions and emulsions is useful in various washing processes. Many industrial processes rely on reactions using chemicals dissolved in water, suspension of solids in water slurries or using water to dissolve and extract substances. Washing is also an important component of several aspects of personal body hygiene.

Transportation

The use of water for transportation of materials through rivers and canals as well as the international shipping lanes is an important part of the world economy.

Chemical Uses

Water is widely used in chemical reactions as a solvent or reactant and less commonly as a solute or catalyst. In inorganic reactions, water is a common solvent, dissolving many ionic compounds. In organic reactions, it is not usually used as a reaction solvent, because it does not dissolve the reactants well and is amphoteric (acidic *and* basic) and nucleophilic. Nevertheless, these properties are sometimes desirable. Also, acceleration of Diels-Alder reactions by water has been observed. Supercritical water has recently been a topic of research. Oxygen-saturated supercritical water combusts organic pollutants efficiently.

Heat Exchange

Water and steam are a common fluid used for heat exchange, due to its availability and high heat capacity, both for cooling and heating. Cool water may even be naturally available from a lake or the sea. It's especially effective to transport heat through vaporization and condensation of water because of its large latent heat of vaporization. A disadvantage is that metals commonly found in industries such as steel and copper are oxidized faster by untreated water and steam. In almost all thermal power stations, water is used as the working fluid (used in a closed loop between boiler, steam turbine and condenser), and the coolant (used to exchange the waste heat to a water body or carry it away by evaporation in a cooling tower). In the United States, cooling power plants is the largest use of water.

In the nuclear power industry, water can also be used as a neutron moderator. In most nuclear reactors, water is both a coolant and a moderator. This provides something of a passive safety measure, as removing the water from the reactor also slows the nuclear reaction down. However other methods are favoured for stopping a reaction and it is preferred to keep the nuclear core covered with water so as to ensure adequate cooling.

Fire Extinction

Water has a high heat of vaporization and is relatively inert, which makes it a good fire extinguishing fluid. The evaporation of water carries heat away from the fire. It is dangerous to use water on fires involving oils and organic solvents, because many organic materials float on water and the water tends to spread the burning liquid. Use of water in fire fighting should also take into account the hazards of a steam explosion, which may occur when water is used on very hot fires in confined spaces, and of a hydrogen explosion, when substances which react with water, such as certain metals or hot carbon such as coal, charcoal, or coke graphite, decompose the water, producing water gas.

The power of such explosions was seen in the Chernobyl disaster, although the water involved did not come from fire-fighting at that time but the reactor's own water cooling system. A steam explosion occurred when the extreme overheating of the core caused water to flash into steam. A hydrogen explosion may have occurred as a result of reaction between steam and hot zirconium.

Recreation

Humans use water for many recreational purposes, as well as for exercising and for sports. Some of these include swimming, waterskiing, boating, surfing and diving. In addition, some sports, like ice hockey and ice skating, are played on ice. Lakesides, beaches and water parks are popular places for people to go to relax and enjoy recreation. Many find the sound and appearance of flowing water to be calming, and fountains and other water features are popular decorations. Some keep fish and other life in aquariums or ponds for show, fun, and companionship. Humans also use water for snow sports i.e. skiing, sledding, snowmobiling or snowboarding, which require the water to be frozen.

Water Industry

The water industry provides drinking water and wastewater services (including sewage treatment) to households and industry. Water supply facilities include water wells, cisterns for rainwater harvesting, water supply networks, and water purification facilities, water tanks, water towers, water pipes including old aqueducts. Atmospheric water generators are in development.

Drinking water is often collected at springs, extracted from artificial borings (wells) in the ground, or pumped from lakes and rivers. Building more wells in adequate places is thus a possible way to produce more water, assuming the aquifers can supply an adequate flow. Other water sources include rainwater collection. Water may require purification for human consumption. This may involve removal of undissolved substances, dissolved substances and harmful microbes. Popular methods are filtering with sand which only removes undissolved material, while chlorination and boiling kill harmful microbes. Distillation does all three functions. More advanced techniques exist, such as reverse osmosis. Desalination of abundant seawater is a more expensive solution used in coastal arid climates.

The distribution of drinking water is done through municipal water systems, tanker delivery or as bottled water. Governments in many countries have programs to distribute water to the needy at no charge.

Reducing usage by using drinking (potable) water only for human consumption is another option. In some cities such as Hong Kong, sea water is extensively used for flushing toilets citywide in order to conserve fresh water resources.

Polluting water may be the biggest single misuse of water; to the extent that a pollutant limits other uses of the water, it becomes a waste of the resource, regardless of benefits to the polluter. Like other types of pollution, this does not enter standard accounting of market costs, being conceived as externalities for which the market cannot account. Thus other people pay the price of water pollution, while the private firms' profits are not redistributed to the local population, victims of this pollution. Pharmaceuticals consumed by humans often end up in the waterways and can have detrimental effects on aquatic life if they bioaccumulate and if they are not biodegradable.

Wastewater facilities are storm sewers and wastewater treatment plants. Another way to remove pollution from surface runoff water is bioswale.

Industrial Applications

Water is used in power generation. Hydroelectricity is electricity obtained from hydropower. Hydroelectric power comes from water driving a water turbine connected to a generator. Hydroelectricity is a low-cost, non-polluting, renewable energy source. The energy is supplied by the motion of water. Typically a dam is constructed on a river, creating an artificial lake behind it. Water flowing out of the lake is forced through turbines that turn generators.

Pressurized water is used in water blasting and water jet cutters. Also, very high pressure water guns are used for precise cutting. It works very well, is relatively safe, and is not harmful to the environment. It is also used in the cooling of machinery to prevent overheating, or prevent saw blades from overheating.

Water is also used in many industrial processes and machines, such as the steam turbine and heat exchanger, in addition to its use as a chemical solvent. Discharge of untreated water from industrial uses is pollution. Pollution includes discharged solutes (chemical pollution) and discharged coolant water (thermal pollution). Industry requires pure water for many applications and utilizes a variety of purification techniques both in water supply and discharge.

Food Processing

Boiling, steaming, and simmering are popular cooking methods that often require immersing food in water or its gaseous state, steam. Water is also used for dishwashing. Water also plays many critical roles within the field of food science. It is important for a food scientist to understand the roles that water plays within food processing to ensure the success of their products.

Solutes such as salts and sugars found in water affect the physical properties of water. The boiling and freezing points of water are affected by solutes, as well as air pressure, which is in turn is affected by altitude. Water boils at lower temperatures with the lower air pressure that occurs at higher elevations. One mole of sucrose (sugar) per kilogram of water raises the boiling point of water by 0.51 °C (32.918 °F), and one mole of salt per kg raises the boiling point by 1.02 °C (33.836 °F); similarly, increasing the number of dissolved particles lowers water's freezing point.

Solutes in water also affect water activity that affects many chemical reactions and the growth of microbes in food. Water activity can be described as a ratio of the vapor pressure of water in a solution to the vapor pressure of pure water. Solutes in water lower water activity—this is important to know because most bacterial growth ceases at low levels of water activity. Not only does microbial growth affect the safety of food, but also the preservation and shelf life of food.

Water hardness is also a critical factor in food processing and may be altered or treated by using a chemical ion exchange system. It can dramatically affect the quality of a product, as well as playing a role in sanitation. Water hardness is classified based on the amounts of removable calcium carbonate salt the water contains per gallon. Water hardness is measured in grains: 0.064 g calcium carbonate is equivalent to one grain of hardness. Water is classified as soft if it contains 1 to 4 grains, medium if it contains 5 to 10 grains and hard if it contains 11 to 20 grains.

The hardness of water also affects its pH balance, which plays a critical role in food processing. For example, hard water prevents successful production of clear beverages. Water hardness also affects sanitation; with increasing hardness, there is a loss of effectiveness for its use as a sanitizer.

According to a report published by the Water Footprint organization in 2010, a single kilogram of beef requires 15 thousand litres of water; however, the authors also make clear that this is a global average and circumstantial factors determine the amount of water used in beef production.

Forest

A '*Forest* is a large area of land covered with trees or other woody vegetation. Hundreds of more precise definitions of "Forest" are used throughout the world, incorporating factors such as tree density, tree height, land use, legal standing and ecological function. According to the widely-used United Nations Food and Agriculture Organization definition, forests covered an area of four billion hectares (15 million square miles) or approximately 30 percent of the world's land area in 2006.

Forests usually provide a diversity of ecosystem services including recycling carbon dioxide into oxygen, acting as a carbon sink, aiding in regulating climate, purify water, mitigating natural hazards such as floods, and serving as a genetic reserve. Forests also serve as a source of lumber and as recreational areas. Conversely, forests may reduce land productivity, incur economic costs, reduce biodiversity, remove water available water for humans and wildlife and act as reservoirs of human and livestock disease.

Definition

Although *forest* is a term of common parlance, there is no universally recognised precise definition, with more than 800 definitions of forest used around the world. Although a forest is usually defined by the presence of trees, under many definitions an area completely lacking trees may still be considered a forest if it grew trees in the past, will grow trees in the future, or was legally designated as a forest regardless of vegetation type.

There are three broad categories of forest definitions in use: administrative, land use, and land cover. Administrative definitions are based primarily upon the legal designations of land, and commonly bare no relationship to the vegetation growing on the land: land that is legally designated as a forest is defined as a forest even if no trees are growing on it. Land Use definitions are based upon the primary purpose that the land serves. For example, a forest may defined as any land that is used primarily for production of timber. Under such a Land Use definition, cleared roads or infrastructure within an area used for forestry, or areas within the region that have been cleared by harvesting, disease or fire are still considered forests even if they contain no trees. Land Cover definitions define forests based upon the type and density of vegetation growing on the land. Such definitions typically define a forest as an area growing trees above some threshold. These thresholds are typically the number of trees per area (density), the area of ground under the tree canopy (canopy cover) or the section of land that is occupied by the cross-section of tree trunks (basal area). Under such Land Cover definitions, and area of land only be defined as forest if it is growing trees. Areas that fail to meet the Land Cover definition may be still included under while immature trees are establishing if they are expected to meet the definition at maturity.

Under land use definitions, there is considerable variation on where the cutoff points are between a forest, woodland, and savanna. Under some definitions, forests require very high levels of tree canopy cover, from 60% to 100%, excluding savannas and woodlands in which trees have a lower canopy cover. Other definitions consider savannas to be a type of forest, and include all areas with tree canopies over 10%.

Distribution

Forests can be found in all regions capable of sustaining tree growth, at altitudes up to the tree line, except where natural fire frequency or other disturbance is too high, or where the environment has been altered by human activity.

The latitudes 10° north and south of the equator are mostly covered in tropical rainforest, and the latitudes between 53°N and 67°N have boreal forest. As a general rule, forests dominated by angiosperms (*broadleaf forests*) are more species-rich than those dominated by gymnosperms (*conifer*, *montane*, or *needleleaf forests*), although exceptions exist.

Forests sometimes contain many tree species only within a small area (as in tropical rain and temperate deciduous forests), or relatively few species over large areas (e.g., taiga and arid montane coniferous forests). Forests are often home to many animal and plant species, and biomass per unit area is high compared to other vegetation communities. Much of this biomass occurs below ground in the root systems and as partially decomposed plant detritus. The woody component of a forest contains lignin, which is relatively slow to decompose compared with other organic materials such as cellulose or carbohydrate.

Forests are differentiated from woodlands by the extent of canopy coverage: in a forest, the branches and the foliage of separate trees often meet or interlock, although there can be gaps of varying sizes within an area referred to as forest. A woodland has a more continuously open canopy, with trees spaced farther apart, which allows more sunlight to penetrate to the ground between them (also see: savanna).

Among the major forested biomes are:

- rain forest (tropical and temperate)
- taiga
- temperate hardwood forest
- tropical dry forest

Types of Forests

Forests can be classified in different ways and to different degrees of specificity. One such way is in terms of the "biome" in which they exist, combined with leaf longevity of the dominant species (whether they are evergreen or deciduous). Another distinction is whether the forests are composed predominantly of broadleaf trees, coniferous (needle-leaved) trees, or mixed.

- Boreal forests occupy the subarctic zone and are generally evergreen and coniferous.
- Temperate zones support both broadleaf deciduous forests (*e.g.*, temperate deciduous forest) and evergreen coniferous forests (*e.g.*, temperate coniferous forests and temperate rainforests). Warm temperate zones support broadleaf evergreen forests, including laurel forests.
- Tropical and subtropical forests include tropical and subtropical moist forests, tropical and subtropical dry forests, and tropical and subtropical coniferous forests.
- Physiognomy classifies forests based on their overall physical structure or developmental stage (e.g. old growth vs. second growth).
- Forests can also be classified more specifically based on the climate and the dominant tree species present, resulting in numerous different forest types (e.g., ponderosa pine/Douglas-fir forest).

A number of global forest classification systems have been proposed, but none has gained universal acceptance. UNEP-WCMC's forest category classification system is a simplification of other more complex systems (e.g. UNESCO's forest and woodland 'subformations'). This system divides the world's forests into 26 major types, which reflect climatic zones as well as the principal types of trees. These 26 major types can be reclassified into 6 broader categories: temperate needleleaf; temperate broadleaf and mixed; tropical moist; tropical dry; sparse trees and parkland; and forest plantations. Each category is described as a separate section below.

Temperate Needleleaf

Temperate needleleaf forests mostly occupy the higher latitude regions of the northern hemisphere, as well as high altitude zones and some warm temperate areas, especially on nutrient-poor or otherwise unfavourable soils. These forests are composed entirely, or nearly so, of coniferous species (Coniferophyta). In the Northern Hemisphere pines *Pinus*, spruces *Picea*, larches *Larix*, firs *Abies*, Douglas firs *Pseudotsuga* and hemlocks *Tsuga*, make up the canopy, but other taxa are also important. In the Southern Hemisphere, most coniferous trees (members of the Araucariaceae and Podocarpaceae) occur in mixtures with broadleaf species, and are classed as broadleaf and mixed forests.

Temperate Broadleaf and Mixed

Temperate broadleaf and mixed forests include a substantial component of trees in the Anthophyta. They are generally characteristic of the warmer temperate latitudes, but extend to cool temperate ones, particularly in the southern hemisphere. They include such forest types as the mixed deciduous forests of the United States and their counterparts in China and Japan, the broadleaf evergreen rainforests of Japan, Chile and Tasmania, the sclerophyllous forests of Australia, central Chile, the Mediterranean and California, and the southern beech Nothofagus forests of Chile and New Zealand.

Tropical Moist

There are many different types of tropical moist forests, although most extensive are the lowland evergreen broad leaf rainforests, for example várzea and igapσ forests and the terra firma forests of the Amazon Basin; the peat swamp forests, dipterocarp forests of Southeast Asia; and the high forests of the Congo Basin. Forests located on mountains are also included in this category, divided largely into upper and lower montane formations on the basis of the variation of physiognomy corresponding to changes in altitude.

Tropical Dry

Tropical dry forests are characteristic of areas in the tropics affected by seasonal drought. The seasonality of rainfall is usually reflected in the deciduousness of the forest canopy, with most trees being leafless for several months of the year. However, under some conditions, e.g. less fertile soils or less predictable drought regimes, the proportion of evergreen species increases and the forests are characterised as "sclerophyllous". Thorn forest, a dense forest of low stature with a high frequency of thorny or spiny species, is found where drought is prolonged, and especially where grazing animals are plentiful. On very poor soils, and especially where fire is a recurrent phenomenon, woody savannas develop.

Sparse Trees and Parkland

Sparse trees and parkland are forests with open canopies of 10–30% crown cover. They occur principally in areas of transition from forested to non-forested landscapes. The two major zones in which these ecosystems occur are in the boreal region and in the seasonally dry tropics. At high latitudes, north of the main zone of boreal forest or taiga, growing conditions are not adequate to maintain a continuous closed forest cover, so tree cover is both sparse and discontinuous. This vegetation is variously called open taiga, open lichen woodland, and forest tundra. It is species-poor, has high bryophyte cover, and is frequently affected by fire.

Forest Plantations

Forest plantations, generally intended for the production of timber and pulpwood increase the total area of forest worldwide. Commonly mono-specific and/or composed of introduced tree species, these ecosystems are not generally important as habitat for native biodiversity. However, they can be managed in ways that enhance their biodiversity protection functions and they are important providers of ecosystem services such as maintaining nutrient capital, protecting watersheds and soil structure as well as storing carbon. They may also play an important role in alleviating pressure on natural forests for timber and fuelwood production.

Forest Categories

28 forest categories are used to enable the translation of forest types from national and regional classification systems to a harmonised global one:

Temperate and Boreal Forest Types

1. Evergreen needleleaf forest – Natural forest with > 30% canopy cover, in which the canopy is predominantly (> 75%) needleleaf and evergreen.
2. Deciduous needleleaf forests – Natural forests with > 30% canopy cover, in which the canopy is predominantly (> 75%) needleleaf and deciduous.
3. Mixed broadleaf/needleleaf forest – Natural forest with > 30% canopy cover, in which the canopy is composed of a more or less even mixture of needleleaf and broadleaf crowns (between 50:50% and 25:75%).
4. Broadleaf evergreen forest – Natural forests with > 30% canopy cover, the canopy being > 75% evergreen and broadleaf.
5. Deciduous broadleaf forest – Natural forests with > 30% canopy cover, in which > 75% of the canopy is deciduous and broadleaves predominate (> 75% of canopy cover).
6. Freshwater swamp forest – Natural forests with > 30% canopy cover, composed of trees with any mixture of leaf type and seasonality, but in which the predominant environmental characteristic is a waterlogged soil.

7. Sclerophyllous dry forest – Natural forest with > 30% canopy cover, in which the canopy is mainly composed of sclerophyllous broadleaves and is > 75% evergreen.
8. Sparse trees and parkland – Natural forests in which the tree canopy cover is between 10–30%, such as in the steppe regions of the world. Trees of any type (e.g., needleleaf, broadleaf, palms).
9. Disturbed natural forest – Any forest type above that has in its interior significant areas of disturbance by people, including clearing, felling for wood extraction, anthropogenic fires, road construction, etc.
10. Exotic species plantation – Intensively managed forests with > 30% canopy cover, which have been planted by people with species not naturally occurring in that country.
11. Native species plantation – Intensively managed forests with > 30% canopy cover, which have been planted by people with species that occur naturally in that country.
12. *Unspecified forest plantation – Forest plantations showing extent only with no further information about their type, This data currently only refers to the Ukraine.
13. *Unclassified forest data – Forest data showing forest extent only with no further information about their type.

Those marked * have been created as a result of data holdings which do not specify the forest type, hence 26 categories are quoted, not 28 shown here.

Tropical Forest Types

1. Lowland evergreen broadleaf rain forest – Natural forests with > 30% canopy cover, below 1,200 m (3,937 ft) altitude that display little or no seasonality, the canopy being >75% evergreen broadleaf.
2. Lower montane forest – Natural forests with > 30% canopy cover, between 1200–1800 m altitude, with any seasonality regime and leaf type mixture.
3. Upper montane forest – Natural forests with > 30% canopy cover, above 1,800 m (5,906 ft) altitude, with any seasonality regime and leaf type mixture.
4. Freshwater swamp forest – Natural forests with > 30% canopy cover, below 1,200 m (3,937 ft) altitude, composed of trees with any mixture of leaf type and seasonality, but in which the predominant environmental characteristic is a waterlogged soil.

5. Semi-evergreen moist broadleaf forest – Natural forests with > 30% canopy cover, below 1,200 m (3,937 ft) altitude in which between 50–75% of the canopy is evergreen, > 75% are broadleaves, and the trees display seasonality of flowering and fruiting.
6. Mixed broadleaf/needleleaf forest – Natural forests with > 30% canopy cover, below 1,200 m (3,937 ft) altitude, in which the canopy is composed of a more or less even mixture of needleleaf and broadleaf crowns (between 50:50% and 25:75%).
7. Needleleaf forest – Natural forest with > 30% canopy cover, below 1,200 m (3,937 ft) altitude, in which the canopy is predominantly (> 75%) needleleaf.
8. Mangroves – Natural forests with > 30% canopy cover, composed of species of mangrove tree, generally along coasts in or near brackish or seawater.
9. Deciduous/semi-deciduous broadleaf forest – Natural forests with > 30% canopy cover, below 1,200 m (3,937 ft) altitude in which between 50–100% of the canopy is deciduous and broadleaves predominate (> 75% of canopy cover).
10. Sclerophyllous dry forest – Natural forests with > 30% canopy cover, below 1,200 m (3,937 ft) altitude, in which the canopy is mainly composed of sclerophyllous broadleaves and is > 75% evergreen.
11. Thorn forest – Natural forests with > 30% canopy cover, below 1,200 m (3,937 ft) altitude, in which the canopy is mainly composed of deciduous trees with thorns and succulent phanerophytes with thorns may be frequent.
12. Sparse trees and parkland – Natural forests in which the tree canopy cover is between 10–30%, such as in the savannah regions of the world. Trees of any type (e.g., needleleaf, broadleaf, palms).
13. Disturbed natural forest – Any forest type above that has in its interior significant areas of disturbance by people, including clearing, felling for wood extraction, anthropogenic fires, road construction, etc.
14. Exotic species plantation – Intensively managed forests with > 30% canopy cover, which have been planted by people with species not naturally occurring in that country.
15. Native species plantation – Intensively managed forests with > 30% canopy cover, which have been planted by people with species that occur naturally in that country.

Forest Loss and Management

The scientific study of forest species and their interaction with the environment is referred to as forest ecology, while the management of forests is often referred to as forestry. Forest management has changed considerably over the last few centuries, with rapid changes from the 1980s onwards culminating in a practice now referred to as sustainable forest management. Forest ecologists concentrate on forest patterns and processes, usually with the aim of elucidating cause-and-effect relationships. Foresters who practice sustainable forest management focus on the integration of ecological, social, and economic values, often in consultation with local communities and other stakeholders.

Anthropogenic factors that can affect forests include logging, urban sprawl, human-caused forest fires, acid rain, invasive species, and the slash and burn practices of swidden agriculture or shifting cultivation. The loss and re-growth of forest leads to a distinction between two broad types of forest, primary or old-growth forest and secondary forest. There are also many natural factors that can cause changes in forests over time including forest fires, insects, diseases, weather, competition between species, etc. In 1997, the World Resources Institute recorded that only 20% of the world's original forests remained in large intact tracts of undisturbed forest. More than 75% of these intact forests lie in three countries—the Boreal forests of Russia and Canada and the rainforest of Brazil.

Canada has about 4,020,000 square kilometres (1,550,000 sq mi) of forest land. More than 90% of forest land is publicly owned and about 50% of the total forest area is allocated for harvesting. These allocated areas are managed using the principles of sustainable forest management, which includes extensive consultation with local stakeholders. About eight percent of Canada's forest is legally protected from resource development (Global Forest Watch Canada) (Natural Resources Canada). Much more forest land—about 40 percent of the total forest land base—is subject to varying degrees of protection through processes such as integrated land use planning or defined management areas such as certified forests (Natural Resources Canada).

These maps represent only virgin forest lost. Some regrowth has occurred but not to the age, size, or extent of 1620, due to population increases and food cultivation. *From William B. Greeley's, The Relation of Geography to Timber Supply, Economic Geography, 1925, vol. 1, p. 1–11. Source of "Today" map: compiled by George Draffan from roadless area map in The Big Outside: A Descriptive Inventory of the*

Big Wilderness Areas of the United States, by Dave Foreman and Howie Wolke (Harmony Books, 1992).'

By December 2006, over 1,237,000 square kilometres of forest land in Canada (about half the global total) had been certified as being sustainably managed (Canadian Sustainable Forestry Certification Coalition). Clearcutting, first used in the latter half of the 20th century, is less expensive, but devastating to the environment, and companies are required by law to ensure that harvested areas are adequately regenerated. Most Canadian provinces have regulations limiting the size of clearcuts, although some older clearcuts can range upwards of 110 square kilometres (27,000 acres) in size which were cut over several years. China instituted a ban on logging, beginning in 1998, due to the erosion and flooding that it caused.

In 2010, the Food and Agriculture Organization of the United Nations reported that world deforestation, mainly the conversion of tropical forests to agricultural land, had decreased over the past ten years but still continues at an alarmingly high rate in many countries. Globally, around 13 million hectares of forests were converted to other uses or lost through natural causes each year between 2000 and 2010 as compared to around 16 million hectares per year during the 1990s. The study covered 233 countries and areas. Brazil and Indonesia, which had the highest loss of forests in the 1990s, have significantly reduced their deforestation rates. In addition, ambitious tree planting programmes in countries such as China, India, the United States and Viet Nam - combined with natural expansion of forests in some regions - have added more than seven million hectares of new forests annually. As a result the net loss of forest area was reduced to 5.2 million hectares per year between 2000 and 2010, down from 8.3 million hectares annually in the 1990s.

In the United States, most forests have historically been affected by humans to some degree, though in recent years improved forestry practices has helped regulate or moderate large scale or severe impacts. However, the United States Forest Service estimates a net loss of about 2 million hectares (4,942,000 acres) between 1997 and 2020; this estimate includes conversion of forest land to other uses, including urban and suburban development, as well as afforestation and natural reversion of abandoned crop and pasture land to forest. However, in many areas of the United States, the area of forest is stable or increasing, particularly in many northern states. The opposite problem from flooding has plagued national forests, with loggers complaining that a lack of thinning and proper forest management has resulted in large forest fires.

Old-growth forest contains mainly natural patterns of biodiversity in established seral patterns, and they contain mainly species native to the region and habitat. The natural formations and processes have not been affected by humans with a frequency or intensity to change the natural structure and components of the habitat. Secondary forest contains significant elements of species which were originally from other regions or habitats.

Smaller areas of woodland in cities may be managed as Urban forestry, sometimes within public parks. These are often created for human benefits; Attention Restoration Theory argues that spending time in nature reduces stress and improves health, while forest schools and kindergartens help young people to develop social as well as scientific skills in forests. These typically need to be close to where the children live, for practical logistics.

Forest Land Area

Forest Land Area						
	2008		*2009*		*2010*	
	('000 km²)	('000 mi²)	('000 km²)	('000 mi²)	('000 km²)	('000 mi²)
Australia	1,511	583	1,502	580	1,493	576
Brazil	5,239	2,023	5,217	2,014	5,195	2,006
Canada	3,101	1,197	3,101	1,197	3,101	1,197
China	2,013	777	2,041	788	2,069	799
European Union	1,559	602	1,564	604	1,569	606
Germany	111	43	111	43	111	43
India	681	263	683	264	684	264
Indonesia	958	370	951	367	944	364
Japan	250	97	250	97	250	97
Russia	8,090	3,120	8,090	3,120	8,091	3,124
United States	3,033	1,171	3,036	1,172	3,040	1,170
World total	40,318	15,567	40,261	15,545	40,204	15,523

Mineral

A mineral is a naturally occurring substance that is solid and inorganic representable by a chemical formula, usually abiogenic, and has an ordered atomic structure. It is different from a rock, which can be an aggregate of minerals or non-minerals and does not have a specific chemical composition. The exact definition of a mineral is under debate, especially with respect to the requirement a valid species be abiogenic, and to a lesser extent with regard to it having an ordered atomic structure. The study of minerals is called mineralogy.

There are over 4,900 known mineral species; over 4,660 of these have been approved by the International Mineralogical Association

(IMA). The silicate minerals compose over 90% of the Earth's crust. The diversity and abundance of mineral species is controlled by the Earth's chemistry. Silicon and oxygen constitute approximately 75% of the Earth's crust, which translates directly into the predominance of silicate minerals. Minerals are distinguished by various chemical and physical properties.

Differences in chemical composition and crystal structure distinguish various species, and these properties in turn are influenced by the mineral's geological environment of formation. Changes in the temperature, pressure, and bulk composition of a rock mass cause changes in its mineralogy; however, a rock can maintain its bulk composition, but as long as temperature and pressure change, its mineralogy can change as well.

Figure: *Amethyst, a variety of quartz*

Minerals can be described by various physical properties which relate to their chemical structure and composition. Common distinguishing characteristics include crystal structure and habit, hardness, lustre, diaphaneity, colour, streak, tenacity, cleavage, fracture, parting, and specific gravity. More specific tests for minerals include reaction to acid, magnetism, taste or smell, and radioactivity.

Minerals are classified by key chemical constituents; the two dominant systems are the Dana classification and the Strunz classification. The silicate class of minerals is subdivided into six subclasses by the degree of polymerization in the chemical structure. All silicate minerals have a base unit of a $[SiO_4]^{4-}$ silica tetrahedra—that is, a silicon cation coordinated by four oxygen anions, which gives the shape of a tetrahedron. These tetrahedra can be polymerized to give the subclasses: orthosilicates (no polymerization, thus single tetrahedra), disilicates (two tetrahedra bonded together), cyclosilicates (rings of tetrahedra), inosilicates (chains of tetrahedra), phyllosilicates (sheets of tetrahedra), and tectosilicates (three-dimensional network of tetrahedra). Other important mineral groups include the native elements, sulfides, oxides, halides, carbonates, sulfates, and phosphates.

Definition

Basic Definition

The general definition of a mineral encompasses the following criteria:

1. Naturally occurring
2. Stable at room temperature
3. Represented by a chemical formula
4. Usually abiogenic (not resulting from the activity of living organisms)
5. Ordered atomic arrangement

The first three general characteristics are less debated than the last two. The first criterion means that a mineral has to form by a natural process, which excludes anthropogenic compounds. Stability at room temperature, in the simplest sense, is synonymous to the mineral being solid. More specifically, a compound has to be stable or metastable at 25 °C. Classical examples of exceptions to this rule include native mercury, which crystallizes at "39 °C, and water ice, which is solid only below 0 °C; as these two minerals were described prior to 1959, they were grandfathered by the International Mineralogical Association (IMA). Modern advances have included extensive study of liquid crystals, which also extensively involve mineralogy. Minerals are chemical compounds, and as such they can be described by fixed or a variable formula. Many mineral groups and species are composed of a solid solution; pure substances are not usually found because of contamination or chemical substitution. For example, the olivine group is described by the variable formula $(Mg, Fe)_2SiO_4$, which is a solid solution of two end-member species, magnesium-rich forsterite and iron-rich fayalite, which are described by a fixed chemical formula. Mineral species themselves could have a variable compositions, such as the sulfide mackinawite, $(Fe, Ni)_9S_8$, which is mostly a ferrous sulfide, but has a very significant nickel impurity that is reflected in its formula.

The requirement of a valid mineral species to be abiogenic has also been described as similar to have to be inorganic; however, this criterion is imprecise and organic compounds have been assigned a separate classification branch. Finally, the requirement of an ordered atomic arrangement is usually synonymous to being crystalline; however, crystals are periodic in addition to being ordered, so the broader criterion is used instead. The presence of an ordered atomic arrangement translates to a variety of macroscopic physical properties, such as crystal form, hardness, and cleavage. There have been several recent proposals to amend the definition to consider biogenic or amorphous substances as minerals. The formal definition of a mineral approved by the IMA in 1995:

> *"A mineral is an element or chemical compound that is normally crystalline and that has been formed as a result of geological processes."*

In addition, biogenic substances were explicitly excluded:

> *"Biogenic substances are chemical compounds produced entirely by biological processes without a geological component (e.g., urinary calculi, oxalate crystals in plant tissues, shells of marine molluscs, etc.) and are not regarded as minerals. However, if geological processes were involved in the genesis of the compound, then the product can be accepted as a mineral."*

Recent Advances

Mineral classification schemes and their definitions are evolving to match recent advances in mineral science. Recent changes have included the addition of an organic class, in both the new Dana and the Strunz classification schemes. The organic class includes a very rare group of minerals with hydrocarbons. The IMA Commission on New Minerals and Mineral Names adopted in 2009 a hierarchical scheme for the naming and classification of mineral groups and group names and established seven commissions and four working groups to review and classify minerals into an official listing of their published names. According to these new rules, "mineral species can be grouped in a number of different ways, on the basis of chemistry, crystal structure, occurrence, association, genetic history, or resource, for example, depending on the purpose to be served by the classification."

The Nickel (1995) exclusion of biogenic substances was not universally adhered to. For example, Lowenstam (1981) stated that "organisms are capable of forming a diverse array of minerals, some of which cannot be formed inorganically in the biosphere." The distinction is a matter of classification and less to do with the constituents of the minerals themselves. Skinner (2005) views all solids as potential minerals and includes biominerals in the mineral kingdom, which are those that are created by the metabolic activities of organisms. Skinner expanded the previous definition of a mineral to classify "element or compound, amorphous or crystalline, formed through *biogeochemical* processes," as a mineral.

Recent advances in high-resolution genetic and x-ray absorption spectroscopy is opening new revelations on the biogeochemical relations between microorganisms and minerals that may make Nickel's (1995) biogenic mineral exclusion obsolete and Skinner's (2005) biogenic mineral inclusion a necessity. For example, the IMA commissioned 'Environmental Mineralogy and Geochemistry Working Group' deals with minerals in the hydrosphere, atmosphere, and biosphere. Mineral forming microorganisms inhabit the areas that this working group deals with. These organisms exist on nearly every rock, soil, and particle surface spanning the globe reaching depths at 1600 metres below the sea floor (possibly further) and 70 kilometres into the stratosphere (possibly entering the mesosphere). Biologists and geologists have started to research and appreciate the magnitude of mineral geoengineering that these creatures are capable of. Bacteria have contributed to the formation of minerals for billions of years and critically define the biogeochemical cycles on this planet. Microorganisms can precipitate metals from solution contributing to the formation of ore deposits in addition to their ability to catalyze mineral dissolution, to respire, precipitate, and form minerals.

Prior to the International Mineralogical Association's listing, over 60 biominerals had been discovered, named, and published. These minerals (a sub-set tabulated in Lowenstam (1981)) are considered minerals proper according to the Skinner (2005) definition. These biominerals are not listed in the International Mineral Association official list of mineral names, however, many of these biomineral representatives are distributed amongst the 78 mineral classes listed in the Dana classification scheme. Another rare class of minerals (primarily biological in origin) include the mineral liquid crystals that are crystalline and liquid at the same time. To date over 80,000 liquid crystalline compounds have been identified.

Concerning the use of the term "mineral" to name this family of liquid crystals, one can argue that the term inorganic would be more appropriate. However, inorganic liquid crystals have long been used for organometallic liquid crystals. Therefore, in order to avoid any confusion between these fairly chemically different families, and taking into account that a large number of these liquid crystals occur naturally in nature, we think that the use of the old fashioned but adequate "mineral" adjective taken sensus largo is more specific that an alternative such as "purely inorganic", to name this subclass of the inorganic liquid crystals family.

The Skinner (2005) definition of a mineral takes this matter into account by stating that a mineral can be crystalline or amorphous. Liquid mineral crystals are amorphous. Biominerals and liquid mineral crystals, however, are not the primary form of minerals, most are geological in origin, but these groups do help to identify at the margins of what constitutes a mineral proper. The formal Nickel (1995) definition explicitly mentioned crystalline nature as a key to defining a substance as a mineral. A 2011 article defined icosahedrite, an aluminium-iron-copper alloy as mineral; named for its unique natural icosahedral symmetry, it is also a quasicrystal. Unlike a true crystal, quasicrystals are ordered but not periodic.

Rocks, Ores, and Gems

Minerals are not equivalent to rocks. Whereas a mineral is a naturally occurring usually solid substance, stable at room temperature, representable by a chemical formula, usually abiogenic, and has an ordered atomic structure, a rock is either an aggregate of one or more minerals, or not composed of minerals at all. Rocks like limestone or quartzite are composed primarily of one mineral—calcite or aragonite in the case of limestone, and quartz in the latter case. Other rocks can be defined by relative abundances of key (essential) minerals; a granite is defined by proportions of quartz, alkali feldspar, and plagioclase feldspar. The other minerals in the rock are termed accessory, and do not greatly affect the bulk composition of the rock. Rocks can also be composed entirely of non-mineral material; coal is a sedimentary rock composed primarily of organically derived carbon.

In rocks, some mineral species and groups are much more abundant than others; these are termed the rock-forming minerals. The major examples of these are quartz, the feldspars, the micas, the amphiboles, the pyroxenes, the olivines, and calcite; except the last one, all of the minerals are silicates. Overall, around 150 minerals are considered particularly important, whether in terms of their abundance or aesthetic value in terms of collecting.

Commercially valuable minerals and rocks are referred to as industrial minerals. For example, muscovite, a white mica, can be used for windows (sometimes referred to as isinglass), as a filler, or as an insulator. Ores are minerals that have a high concentration of a certain element, typically a metal. Examples are cinnabar (HgS), an ore of mercury, sphalerite (ZnS), an ore of zinc, or cassiterite (SnO_2), an ore of tin. Gems are minerals with an ornamental value, and are distinguished from non-gems by their beauty, durability, and usually, rarity. There are about 20 mineral species that qualify as gem minerals, which constitute about 35 of the most common gemstones. Gem minerals are often present in several varieties, and so one mineral can account for several different gemstones; for example, ruby and sapphire are both corundum, Al_2O_3.

Nomenclature and Classification

In general, a mineral is defined as naturally occurring solid, that is stable at room temperature, representable by a chemical formula, usually abiogenic, and has an ordered atomic structure. However, a mineral can be also narrowed down in terms of a mineral group, series, species, or variety, in order from most broad to least broad. The basic level of definition is that of mineral species, which is distinguished from other species by specific and unique chemical and physical properties. For example, quartz is defined by its formula, SiO_2, and a specific crystalline structure that distinguishes it from other minerals with the same chemical formula (termed polymorphs). When there exists a range of composition between two minerals species, a mineral series is defined. For example, the biotite series is represented by variable amounts of the endmembers phlogopite, siderophyllite, annite, and eastonite. In contrast, a mineral group is a grouping of mineral species with some common chemical properties that share a crystal structure. The pyroxene group has a common formula of $XY(Si,Al)_2O_6$, where X and Y are both cations, with X typically bigger than Y; the pyroxenes are single-chain silicates that crystallize in either the orthorhombic or monoclinic crystal systems. Finally, a mineral variety is a specific type of mineral species that differs by some physical characteristic, such as colour or crystal habit. An example is amethyst, which is a purple variety of quartz.

Two common classifications are used for minerals; both the Dana and Strunz classifications rely on the composition of the mineral, specifically with regards to important chemical groups, and its structure. The Dana *System of Mineralogy* was first published in 1837 by James Dwight Dana, a leading geologist of his time; it is in its eighth edition (1997 ed.). The Dana classification, assigns a four-part number to a mineral species. First is its class, based on important compositional groups; next, the type gives the ratio of cations to anions in the mineral; finally, the last two numbers group minerals by structural similarity with a given type or class. The less commonly used Strunz classification, named for German mineralogist Karl Hugo Strunz, is based on the Dana system, but combines both chemical and structural criteria, the latter with regards to distribution of chemical bonds.

There are over 4,660 approved mineral species. They are most commonly named after a person (45%), followed by discovery location (23%); names based on chemical composition (14%) and physical properties (8%) are the two other major groups of mineral name etymologies. The common suffix *-ite* of mineral names descends from the ancient Greek suffix - ß τ η ò (-ites), meaning "connected with or belonging to".

Mineral Chemistry

The abundance and diversity of minerals is controlled directly by their chemistry, in turn dependent on elemental abundances in the Earth. The majority of minerals observed are derived from the Earth's crust. Eight elements account for most of the key components of minerals, due to their abundance in the crust. These eight elements, summing to over 98% of the crust by weight, are, in order of decreasing abundance: oxygen, silicon, aluminium, iron, magnesium, calcium, sodium and potassium. Oxygen and silicon are by far the two most important — oxygen composes 46.6% of the crust by weight, and silicon accounts for 27.7%.

The minerals that form are directly controlled by the bulk chemistry of the parent body. For example, a magma rich in iron and magnesium will form mafic minerals, such as olivine and the pyroxenes; in contrast, a more silica-rich magma will crystallize to form minerals than incorporate more SiO_2, such as the feldspars and quartz. In a limestone, calcite or aragonite (both $CaCO_3$) form because the rock is rich in calcium and carbonate. A corollary is that a mineral will not be found in a rock whose bulk chemistry does not resemble the bulk chemistry of a given mineral with the exception of trace minerals. For example, kyanite, Al_2SiO_5 forms from the metamorphism of aluminium-rich shales; it would not likely occur in aluminium-poor rock, such quartzite.

The chemical composition may vary between end member species of a mineral series. For example, the plagioclase feldspars comprise a continuous series from sodium-rich end member albite ($NaAlSi_3O_8$) to calcium-rich anorthite ($CaAl_2Si_2O_8$) with four recognized intermediate varieties between them (given in order from sodium- to calcium-rich): oligoclase, andesine, labradorite, and bytownite. Other examples of series include the olivine series of magnesium-rich forsterite and iron-rich fayalite, and the wolframite series of manganese-rich hübnerite and iron-rich ferberite.

Chemical substitution and coordination polyhedra explain this common feature of minerals. In nature, minerals are not pure substances, and are contaminated by whatever other elements are present in the given chemical system. As a result, it is possible for one element to be substituted for another. Chemical substitution will occur between ions of a similar size and charge; for example, K^+ will not substitute for Si^{4+} because of chemical and structural incompatibilities caused by a big difference in size and charge.

A common example of chemical substitution is that of Si^{4+} by Al^{3+}, which are close in charge, size, and abundance in the crust. In the example of plagioclase, there are three cases of substitution. Feldspars are all framework silicates, which have a silicon-oxygen ratio of 2:1, and the space for other elements is given by the substitution of Si^{4+} by Al^{3+} to give a base unit of $[AlSi_3O_8]^-$; without the substitution, the formula would be charge-balanced as SiO_2, giving quartz. The significance of this structural property will be explained further by coordination polyhedra. The second substitution occurs between Na^+ and Ca^{2+}; however, the difference in charge has to accounted for by making a second substitution of Si^{4+} by Al^{3+}.

Coordination polyhedra are geometric representation of how a cation is surrounded by an anion. In mineralogy, due its abundance in the crust, coordination polyhedra are usually considered in terms of oxygen. The base unit of silicate minerals is the silica tetrahedron — one Si^{4+} surrounded by four O^{2-}.

An alternate way of describing the coordination of the silicate is by a number: in the case of the silica tetrahedron, the silicon is said to have a coordination number of 4. Various cations have a specific range of possible coordination numbers; for silicon, it is almost always 4, except for very high-pressure minerals where compound is compressed such that silicon is in six-fold (octahedral) coordination by oxygen. Bigger cations have a bigger coordination number because of the increase in relative size as compared to oxygen (the last orbital subshell of heavier atoms is different too). Changes in coordination numbers between leads to physical and mineralogical differences; for example, at high pressure such as in the mantle, many minerals, especially silicates such as olivine and garnet will change to a perovskite structure, where silicon is in octahedral coordination.

Another example are the aluminosilicates kyanite, andalusite, and sillimanite (polymorphs, as they share the formula Al_2SiO_5), which differ by the coordination number of the Al^{3+}; these minerals transition from one another as a response to changes in pressure and temperature. In the case of silicate materials, the substitution of Si^{4+} by Al^{3+} allows for a variety of minerals because of the need to balance charges.

Changes in temperature and pressure, and composition alter the mineralogy of a rock sample. Changes in composition can be caused by processes such as weathering or metasomatism (hydrothermal alteration). Changes in temperature and pressure occur when the host rock undergoes tectonic or magmatic movement into differing physical regimes.

Changes in thermodynamic conditions make it favourable for mineral assemblages to react with each other to produce new minerals; as such, it is possible for two rocks to have an identical or a very similar bulk rock chemistry without having a similar mineralogy. This process of mineralogical alteration is related to the rock cycle. An example of a series of mineral reactions is illustrated as follows.

Orthoclase feldspar ($KAlSi_3O_8$) is a mineral commonly found in granite, a plutonic igneous rock. When exposed to weathering, it reacts to form kaolinite ($Al_2Si_2O_5(OH)_4$, a sedimentary mineral, and silicic acid):

$$2\ KAlSi_3O_8 + 5\ H_2O + 2\ H^+ \rightarrow Al_2Si_2O_5(OH)_4 + 4\ H_2SiO_3 + 2\ K^+$$

Under low-grade metamorphic conditions, kaolinite reacts with quartz to form pyrophyllite ($Al_2Si_4O_{10}(OH)_2$):

$$Al_2Si_2O_5(OH)_4 + SiO_2 \rightarrow Al_2Si_4O_{10}(OH)_2 + H_2O$$

As metamorphic grade increases, the pyrophyllite reacts to form kyanite and quartz:

$$Al_2Si_4O_{10}(OH)_2 \rightarrow Al_2SiO_5 + 3\ SiO_2 + H_2O$$

Alternatively, a mineral may change its crystal structure as a consequence of changes in temperature and pressure without reacting. For example, quartz will change into a variety of its SiO_2 polymorphs, such as tridymite and cristobalite at high temperatures, and coesite at high pressures.

Physical Properties of Minerals

Classifying minerals ranges from simple to difficult. A mineral can be identified by several physical properties, some of them being sufficient for full identification without equivocation. In other cases, minerals can only be classified by more complex optical, chemical or X-ray diffraction analysis; these methods, however, can be costly and time-consuming. Physical properties applied for classification include crystal structure and habit, hardness, lustre, diaphaneity, colour, streak, cleavage and fracture, and specific gravity. Other less general tests include fluorescence, phosphorescence, magnetism, radioactivity, tenacity (response to mechanical induced changes of shape or form), piezoelectricity and reactivity to dilute acids.

Crystal Structure and Habit

Crystal structure results from the orderly geometric spatial arrangement of atoms in the internal structure of a mineral. This crystal structure is based on regular internal atomic or ionic arrangement that is often expressed in the geometric form that the crystal takes. Even when the mineral grains are too small to see or are irregularly shaped, the underlying crystal structure is always periodic and can be determined by X-ray diffraction. Minerals are typically described by their symmetry content. Crystals are restricted to 32 point groups, which differ by their symmetry. These groups are classified in turn into more broad categories, the most encompassing of these being the six crystal families.

These families can be described by the relative lengths of the three crystallographic axes, and the angles between them; these relationships correspond to the symmetry operations that define the narrower point groups. They are summarized below; a, b, and c represent the axes, and α, β, γ represent the angle opposite the respective crystallographic axis (e.g. α is the angle opposite the a-axis, viz. the angle between the b and c axes):

Crystal family	*Lengths*	*Angles*	*Common examples*
Isometric	a=b=c	α=β=γ=90°	Garnet, halite, pyrite
Tetragonal	a=b≠c	α=β=γ=90°	Rutile, zircon, andalusite
Orthorhombic	a≠b≠c	α=β=γ=90°	Olivine, aragonite, orthopyroxenes
Hexagonal	a=b≠c	α=β=90°, γ=120°	Quartz, calcite, tourmaline
Monoclinic	a≠b≠c	α=γ=90°, β≠90°	Clinopyroxenes, orthoclase, gypsum
Triclinic	a≠b≠c	α≠β≠γ≠90°	Anorthite, albite, kyanite

The hexagonal crystal family is also split into two crystal *systems* — the trigonal, which has a three-fold axis of symmetry, and the hexagonal, which has a six-fold axis of symmetry.

Chemistry and crystal structure together define a mineral. With a restriction to 32 point groups, minerals of different chemistry may have identical crystal structure. For example, halite (NaCl), galena (PbS), and periclase (MgO) all belong to the hexaoctahedral point group (isometric family), as they have a similar stoichiometry between their different constituent elements. In contrast, polymorphs are groupings of minerals that share a chemical formula but have a different structure. For example, pyrite and marcasite, both iron sulfides, have the formula FeS_2; however, the former is isometric while the latter is orthorhombic. This polymorphism extends to other sulfides with the generic AX_2 formula; these two groups are collectively known as the pyrite and marcasite groups.

Polymorphism can extend beyond pure symmetry content. The aluminosilicates are a group of three minerals — kyanite, andalusite, and sillimanite — which share the chemical formula Al_2SiO_5. Kyanite is triclinic, while andalusite and sillimanite are both orthorhombic and belong to the dipyramidal point group. These difference arise correspond to how aluminium is coordinated within the crystal structure. In all minerals, one aluminium ion is always in six-fold coordination by oxygen; the silicon, as a general rule is in four-fold coordination in all minerals; an exception is a case like stishovite (SiO_2, an ultra-high pressure quartz polymorph with rutile structure). In kyanite, the second aluminium is in six-fold coordination; its chemical formula can be expressed as $Al^{[6]}Al^{[6]}SiO_5$, to reflect its crystal structure. Andalusite has the second aluminium in five-fold coordination ($Al^{[6]}Al^{[5]}SiO_5$) and sillimanite has it in four-fold coordination ($Al^{[6]}Al^{[4]}SiO_5$).

Differences in crystal structure and chemistry greatly influence other physical properties of the mineral. The carbon allotropes diamond and graphite have vastly different properties; diamond is the hardest natural substance, has an adamantine lustre, and belongs to the isometric crystal family, whereas as graphite is very soft, has a greasy lustre, and crystallises in the hexagonal family. This difference is accounted by differences in bonding. In diamond, the carbons are in sp^3 hybrid orbitals, which means they form a framework where each carbon is covalently bonded to three neighbours in a tetrahedral fashion; on the other hand, graphite is composed of sheets of carbons in sp^2 hybrid orbitals, where each carbon is bonded covalently to only two others. These sheets are held together by much weaker van der Waals forces, and this discrepancy translates to big macroscopic differences. Twinning is the intergrowth of two or more crystal of a single mineral species. The geometry of the twinning is controlled by the mineral's symmetry. As a result, there are several types of twins, including contact twins, reticulated twins, geniculated twins, penetration twins, cyclic twins, and polysynthetic twins.

Contact, or simple twins, consist of two crystals joined at a plane; this type of twinning is common in spinel. Reticulated twins, common in rutile, are interlocking crystals resembling netting. Geniculated twins have a bend in the middle that is caused by start of the twin. Penetration twins consist of two single crystals that have grown into each other; examples of this twinning include cross-shaped staurolite twins and Carlsbad twinning in orthoclase. Cyclic twins are caused by repeated twinning around a rotation axis. It occurs around three, four, five, six, or eight-fold axes, and the corresponding patterns are called threelings, fourlings, fivelings, sixlings, and eightlings. Sixlings are common in aragonite. Polysynthetic twins are similar to cyclic twinning by the presence of repetitive twinning; however, instead of occurring around a rotational axis, it occurs along parallel planes, usually on a microscopic scale.

Crystal habit refers to the overall shape of crystal. Several terms are used to describe this property. Common habits include acicular, which described needlelike crystals like in natrolite, bladed, dendritic (tree-pattern, common in native copper), equant, which is typical of garnet, prismatic (elongated in one direction), and tabular, which differs from bladed habit in that the former is platy whereas the latter has a defined elongation. Related to crystal form, the quality of crystal faces is diagnostic of some minerals, especially with a petrographic microscope. Euhedral crystals have a defined external shape, while anhedral crystals do not; those intermediate forms are termed subhedral.

Hardness

The hardness of a mineral defines how much it can resist scratching. This physical property is controlled by the chemical composition and crystalline structure of a mineral. A mineral's hardness is not necessarily constant for all sides, which is a function of its structure; crystallographic weakness renders some directions softer than others. An example of this property exists in kyanite, which has a Mohs hardness of 5½ parallel to [001] but 7 parallel to [100].

The most common scale of measurement is the ordinal Mohs hardness scale. Defined by ten indicators, a mineral with a higher index scratches those below it. The scale ranges from talc, a phyllosilicate, to diamond, a carbon polymorph that is the hardest natural material. The scale is provided below:

Mohs hardness	***Mineral***	***Chemical formula***
1	Talc	$Mg_3Si_4O_{10}(OH)_2$
2	Gypsum	$CaSO_4 \cdot 2H_2O$
3	Calcite	$CaCO_3$
4	Fluorite	CaF_2
5	Apatite	$Ca_5(PO_4)_3(OH,Cl,F)$
6	Orthoclase	$KAlSi_3O_8$
7	Quartz	SiO_2
8	Topaz	$Al_2SiO_4(OH,F)_2$
9	Corundum	Al_2O_3
10	Diamond	C

Lustre and Diaphaneity

Lustre indicates how light reflects from the mineral's surface, with regards to its quality and intensity. There are numerous qualitative terms used to describe this property, which are split into metallic and non-metallic categories. Metallic and sub-metallic minerals have high reflectivity like metal; examples of minerals with this lustre are galena and pyrite. Non-metallic lustres include: adamantine, such as in diamond; vitreous, which is a glassy lustre very common in silicate minerals; pearly, such as in talc and apophyllite, resinous, such as members of the garnet group, silky which common in fibrous minerals such as asbestiform chrysotile.

The diaphaneity of a mineral describes the ability of light to pass through it. Transparent minerals do not diminish the intensity of light passing through it. An example of such a mineral is muscovite (potassium mica); some varieties are sufficiently clear to have been used for windows. Translucent minerals allow some light to pass, but less than those that are transparent. Jadeite and nephrite (mineral forms of jade are examples of minerals with this property). Minerals that do not allow light to pass are called opaque.

The diaphaneity of a mineral depends on thickness of the sample. When a mineral is sufficiently thin (e.g., in a thin section for petrography), it may become transparent even if that property is not seen in hand sample. In contrast, some minerals, such as hematite or pyrite are opaque even in thin-section.

Colour and Streak

Colour is the most obvious property of a mineral, but it is often non-diagnostic. It is caused by electromagnetic radiation interacting with electrons (except in the case of incandescence, which does not apply to minerals). Two broad classes of elements are defined with regards to their contribution to a mineral's colour. Idiochromatic elements are essential to a mineral's composition; their contribution to a mineral's colour is diagnostic. Examples of such minerals are malachite (green) and azurite (blue). In contrast, allochromatic elements in minerals are present in trace amounts as impurities. An example of such a mineral would be the ruby and sapphire varieties of the mineral corundum. The colours of pseudochromatic minerals are the result of interference of light waves. Examples include opal, labradorite, ammolite and bornite.

In addition to simple body colour, minerals can have various other distinctive optical properties, such as play of colours, asterism, chatoyancy, iridescence, tarnish, and pleochroism. Several of these properties involve variability in colour. Play of colour, such as in opal, results in the sample reflecting different colours as it is turned, while pleochroism describes the change in colour as light passes through a mineral in a different orientation. Iridescence is a variety of the play of colours where light scatters off a coating on the surface of crystal, cleavage planes, or off layers having minor gradations in chemistry. In contrast, the play of colours in opal is caused by light refracting from ordered microscopic silica spheres within its physical structure. Chatoyancy ("cat's eye") is the wavy banding of colour that is observed as the sample is rotated; asterism, a variety of chatoyancy, gives the appearance of a star on the mineral grain. The latter property is particularly common in gem-quality corundum.

The streak of a mineral refers to the colour of a mineral in powdered form, which may or may not be identical to its body colour. The most common way of testing this property is done with a streak plate, which is made out of porcelain and coloured either white or black. The streak of a mineral is independent of trace elements or any weathering surface. A common example of this property is illustrated with hematite, which is coloured black, silver, or red in hand sample, but has a cherry-red to reddish-brown streak. Streak is more often distinctive for metallic minerals, in contrast to non-metallic minerals whose body colour is created by allochromatic elements. Streak testing is constrained by the hardness of the mineral, as those harder than 7 powder the *streak plate* instead.

Cleavage, Parting, Fracture, and Tenacity

By definition, minerals have a characteristic atomic arrangement. Weakness in this crystalline structure causes planes of weakness, and the breakage of a mineral along such planes is termed cleavage. The quality of cleavage can be described based on how cleanly and easily the mineral breaks; common descriptors, in order of decreasing quality, are "perfect", "good", "distinct", and "poor". In particularly transparent mineral, or in thin-section, cleavage can be seen a series of parallel lines marking the planar surfaces when viewed at a side. Cleavage is not a universal property among minerals; for example, quartz, consisting of extensively interconnected silica tetrahedra, does not have a crystallographic weakness which would allow it to cleave. In contrast, micas, which have perfect basal cleavage, consist of sheets of silica tetrahedra which are very weakly held together.

As cleavage is a function of crystallography, there are a variety of cleavage types. Cleavage occurs typically in either one, two, three, four, or six directions. Basal cleavage in one direction is a distinctive property of the micas. Two-directional cleavage is described as prismatic, and occurs in minerals such as the amphiboles and pyroxenes. Minerals such as galena or halite have cubic (or isometric) cleavage in three directions, at 90°; when three directions of cleavage are present, but not at 90°, such as in calcite or rhodochrosite, it is termed rhombohedral cleavage. Octahedral cleavage (four directions) is present in fluorite and diamond, and sphalerite has six-directional dodecahedral cleavage.

Minerals with many cleavages might not break equally well in all of the directions; for example, calcite has good cleavage in three direction, but gypsum has perfect cleavage in one direction, and poor cleavage in two other directions. Angles between cleavage planes vary between minerals. For example, as the amphiboles are double-chain silicates and the pyroxenes are single-chain silicates, the angle between their cleavage planes is different. The pyroxenes cleave in two directions at approximately 90°, whereas the amphiboles distinctively cleave in two directions separated by approximately 120° and 60°. The cleavage angles can be measured with a contact goniometre, which is similar to a protractor.

Parting, sometimes called "false cleavage", is similar in appearance to cleavage but is instead produced by structural defects in the mineral as opposed to systematic weakness. Parting varies from crystal to crystal of a mineral, whereas all crystals of a given mineral will cleave if the atomic structure allows for that property. In general, parting is caused by some stress applied to a crystal. The sources of the stresses include deformation (e.g. an increase in pressure), exsolution, or twinning. Minerals that often display parting include the pyroxenes, hematite, magnetite, and corundum.

When a mineral is broken in a direction that does not correspond to a plane of cleavage, it is termed to have been fractured. There are several types of uneven fracture. The classic example is conchoidal fracture, like that of quartz; rounded surfaces are created, which are marked by smooth curved lines.

This type of fracture occurs only in very homogeneous minerals. Other types of fracture are fibrous, splintery, and hackly. The latter describes a break along a rough, jagged surface; an example of this property is found in native copper.

Tenacity is related to both cleavage and fracture. Whereas fracture and cleavage describes the surfaces that are created when a mineral is broken, tenacity describes how resistant a mineral is to such breaking. Minerals can be described as brittle, ductile, malleable, sectile, flexible, or elastic.

Specific Gravity

Specific gravity numerically describes the density of a mineral. The dimensions of density are mass divided by volume with units: kg/ m^3 or g/cm^3. Specific gravity measures how much water a mineral sample displaces. Defined as the quotient of the mass of the sample and difference between the weight of the sample in air and its corresponding weight in water, specific gravity is a unitless ratio. Among most minerals, this property is not diagnostic. Rock forming minerals — typically silicates or occasionally carbonates — have a specific gravity of 2.5–3.5.

High specific gravity is a diagnostic property of a mineral. A variation in chemistry (and consequently, mineral class) correlates to a change in specific gravity. Among more common minerals, oxides and sulfides tend to have a higher specific gravity as they include elements with higher atomic mass. A generalization is that minerals with metallic or adamantine lustre tend to have higher specific gravities than those having a non-metallic to dull lustre. For example, hematite, Fe_2O_3, has a specific gravity of 5.26 while galena, PbS, has a specific gravity of 7.2–7.6, which is a result of their high iron and lead content, respectively. A very high specific gravity becomes very pronounced in native metals; kamacite, an iron-nickel alloy common in iron meteorites has a specific gravity of 7.9, and gold has an observed specific gravity between 15 and 19.3.

Other Properties

Other properties can be used to diagnose minerals. These are less general, and apply to specific minerals.

Dropping dilute acid (often 10% HCl) aids in distinguishing carbonates from other mineral classes. The acid reacts with the carbonate ($[CO_3]^{2-}$) group, which causes the affected area to effervesce, giving off carbon dioxide gas. This test can be further expanded to test the mineral in its original crystal form or powdered. An example of this test is done when distinguish calcite from dolomite, especially within rocks (limestone and dolostone respectively). Calcite immediately effervesces in acid, whereas acid must be applied to powdered dolomite (often to a scratched surface in a rock), for it to effervesce. Zeolite minerals will not effervesce in acid; instead, they become frosted after 5–10 minutes, and if left in acid for a day, they dissolve or become a silica gel.

When tested, magnetism is a very conspicuous property of minerals. Among common minerals, magnetite exhibits this property strongly, and it is also present, albeit not as strongly, in pyrrhotite and ilmenite.

Minerals can also be tested for taste or smell. Halite, NaCl, is table salt; its potassium-bearing counterpart, sylvite, has a pronounced bitter taste. Sulfides have a characteristic smell, especially as samples are fractured, reacting, or powdered.

Radioactivity is a rare property; minerals may be composed of radioactive elements. They could be a defining constituent, such as uranium in uraninite, autunite, and carnotite, or as trace impurities. In the latter case, the decay of a radioactive element damages the mineral crystal; the result, termed a *radioactive halo* or *pleochroic halo,* is observable by various techniques, such as thin-section petrography.

Mineral Classes

As the composition of the Earth's crust is dominated by silicon and oxygen, silicate elements are by far the most important class of minerals in terms of rock formation and diversity. However, non-silicate minerals are of great economic importance, especially as ores.

Non-silicate minerals are subdivided into several other classes by their dominant chemistry, which included native elements, sulfides, halides, oxides and hydroxides, carbonates and nitrates, borates, sulfates, phosphates, and organic compounds. The majority of non-silicate mineral species are extremely rare (constituting in total 8% of the Earth's crust), although some are relative common, such as calcite, pyrite, magnetite, and hematite. There are two major structural styles observed in non-silicates: close-packing and silicate-like linked tetrahedra. The close-packed structures, which is a way to densely pack atoms while minimizing interstitial space. Hexagonal close-packing involves stacking layers where every other layer is the same ("ababab"), whereas cubic close-packing involves stacking groups of three layers ("abcabcabc"). Analogues to linked silica tetrahedra include SO_4 (sulfate), PO_4 (phosphate), AsO_4 (arsenate), and VO_4 (vanadate). The non-silicates have great economic importance, as they concentrate elements more than the silicate minerals do.

The largest grouping of minerals by far are the silicates; most rocks are composed of greater than 95% silicate minerals, and over 90% of the Earth's crust is composed of these minerals. The two main constituents of silicates are silicon and oxygen, which are the two most abundant elements in the Earth's crust. Other common elements in silicate minerals correspond to other common elements in the Earth's crust, such aluminium, magnesium, iron, calcium, sodium, and potassium. Some important rock-forming silicates include the feldspars, quartz, olivines, pyroxenes, amphiboles, garnets, and micas.

Silicates

The base of unit of a silicate mineral is the $[SiO_4]^{4-}$ tetrahedron. In the vast majority of cases, silicon is in four-fold or tetrahedral coordination with oxygen. In very high-pressure situations, silicon will be six-fold or octahedral coordination, such as in the perovskite structure or the quartz polymorph stishovite (SiO_2). In the latter case, the mineral no longer has a silicate structure, but that of rutile (TiO_2), and its associated group, which are simple oxides. These silica tetrahedra are then polymerized to some degree to create various structures, such as one-dimensional chains, two-dimensional sheets, and three-dimensional frameworks. The basic silicate mineral where no polymerization of the tetrahedra has occurred requires other elements to balance out the base 4- charge. In other silicate structures, different combinations of elements are required to balance out the resultant negative charge. It is common for the Si^{4+} to be substituted by Al^{3+} because of similarity in ionic radius and charge; in those case, the $[AlO_4]^{5-}$ tetrahedra form the same structures as do the unsubstituted tetrahedra, but their charge-balancing requirements are different.

The degree of polymerization can be described by both the structure formed and how many tetrahedral corners (or coordinating oxygens) are shared (for aluminium and silicon in tetrahedral sites). Orthosilicates (or nesosilicates) have no linking of polyhedra, thus tetrahedra share no corners. Disilicates (or sorosilicates) have two tetrahedra sharing one oxygen atom. Inosilicates are chain silicates; single-chain silicates have two shared corners, whereas double-chain silicates have two or three shared corners. In phyllosilicates, a sheet structure is formed which requires three shared oxygens; in the case of double-chain silicates, some tetrahedra must share two corners instead of three as otherwise a sheet structure would result. Framework silicates, or tectosilicates, have tetrahedra that share all four corners. The ring silicates, or cyclosilicates, only need tetrahedra to share two corners to form the cyclical structure.

The silicate subclasses are described below in order of decreasing polymerization.

Tectosilicates

Tectosilicates, also known as framework silicates, have the highest degree of polymerization. With all corners of a tetrahedra shared, the silicon:oxygen ratio becomes 1:2. Examples are quartz, the feldspars, feldspathoids, and the zeolites. Framework silicates tend to be particularly chemically stable as a result of strong covalent bonds.

Forming 12% of the Earth's crust, quartz (SiO_2) is the most abundant mineral species. It is characterized by its high chemical and physical resistivity. Quartz has several polymorphs, including tridymite and cristobalite at high temperatures, high-pressure coesite, and ultra-high pressure stishovite. The latter mineral can only be formed on Earth by meteorite impacts, and its structure has been composed so much that it had changed from a silicate structure to that of rutile (TiO_2). The silica polymorph that is most stable at the Earth's surface is α-quartz. Its counterpart, β-quartz, is present only at high temperatures and pressures (changes to α-quartz below 573 °C at 1 bar). These two polymorphs differ by a "kinking" of bonds; this change in structure gives β-quartz greater symmetry than α-quartz, and they are thus also called high quartz (β) and low quartz (α).

Feldspars are the most abundant group in the Earth's crust, at about 50%. In the feldspars, Al^{3+} substitutes for Si^{4+}, which creates a charge imbalance that must be accounted for by the addition of cations. The base structure becomes either $[AlSi_3O_8]^-$ or $[Al_2Si_2O_8]^{2-}$ There are 22 mineral species of feldspars, subdivided into two major subgroups—alkali and plagioclase—and two less common groups—celsian and banalsite. The alkali feldspars are most commonly in a series between potassium-rich orthoclase and sodium-rich albite; in the case of plagioclase, the most common series ranges from albite to calcium-rich anorthite. Crystal twinning is common in feldspars, especially polysynthetic twins in plagioclase and Carlsbad twins in alkali feldspars. If the latter subgroup cools slowly from a melt, it forms exsolution lamellae because the two components—orthoclase and albite—are unstable in solid solution. Exsolution can be on a scale from microscopic to readily observable in hand-sample; perthitic texture forms when Na-rich feldspar exsolve in a K-rich host. The opposite texture (antiperthitic), where K-rich feldspar exsolves in a Na-rich host, is very rare.

Feldsapthoids are structurally similar to feldspar, but differ in that they form in Si-deficient conditions which allows for further substitution by Al^{3+}. As a result, feldsapthoids cannot be associated with quartz. A common example of a feldsapthoid is nepheline ((Na, K)$AlSiO_4$); compared to alkali feldspar, nepheline has an Al_2O_3:SiO_2 ratio of 1:2, as opposed to 1:6 in the feldspar. Zeolites often have distinctive crystal habits, occurring in needles, plates, or blocky masses. They form in the presence of water at low temperatures and pressures, and have channels and voids in their structure. Zeolites have several industrial applications, especially in waste water treatment.

Phyllosilicates

Phyllosilicates consist of sheets of polymerized tetrahedra. They are bound at three oxygen sites, which gives a characteristic silicon:oxygen ratio of 2:5. Important examples include the mica, chlorite, and the kaolinite-serpentine groups. The sheets are weakly bound by van der Waals forces or hydrogen bonds, which causes a crystallographic weakness, in turn leading to a prominent basal cleavage among the phyllosilicates. In addition to the tetrahedra, phyllosilicates have a sheet of octahedra (elements in six-fold coordination by oxygen) that balanced out the basic tetrahedra, which have a negative charge (e.g. $[Si_4O_{10}]^{4-}$) These tetrahedra (T) and octahedra (O) sheets are stacked in a variety of combinations to create phyllosilicate groups. Within an octahedral sheet, there are three octahedral sites in a unit structure; however, not all of the sites may be occupied. In that case, the mineral is termed dioctahedral, whereas in other case it is termed trioctahedral.

The kaolinite-serpentine group consists of T-O stacks (the 1:1 clay minerals); their hardness ranges from 2 to 4, as the sheets are held by hydrogen bonds. The 2:1 clay minerals (pyrophyllite-talc) consist of T-O-T stacks, but they are softer (hardness from 1 to 2), as they are instead held together by van der Waals forces. These two groups of minerals are subgrouped by octahedral occupation; specifically, kaolinite and pyrophyllite are dioctahedral whereas serpentine and talc trioctahedral.

Micas are also T-O-T-stacked phyllosilicates, but differ from the other T-O-T and T-O-stacked subclass members in that they incorporate aluminium into the tetrahedral sheets (clay minerals have Al^{3+} in octahedral sites). Common examples of micas are muscovite, and the biotite series. The chlorite group is related to mica group, but a brucite-like ($Mg(OH)_2$) layer between the T-O-T stacks.

Because of their chemical structure, phyllosilicates typically have flexible, elastic, transparent layers that are electrical insulators and can be split into very thin flakes. Micas can be used in electronics as insulators, in construction, as optical filler, or even cosmetics. Chrysotile, a species of serpentine, is the most common mineral species in industrial asbestos, as it is less dangerous in terms of health than the amphibole asbestos.

Inosilicates

Inosilicates consist of tetrahedra repeatedly bonded in chains. These chains can be single, where a tetrahedron is bound to two others to form a continuous chain; alternatively, two chains can be merged to create double-chain silicates. Single-chain silicates have a silicon:oxygen ratio of 1:3 (e.g. $[Si_2O_6]^{4-}$), whereas the double-chain variety has a ratio of 4:11, e.g. $[Si_8O_{22}]^{12-}$. Inosilicates contain two important rock-forming mineral groups; single-chain silicates are most commonly pyroxenes, while double-chain silicates are often amphiboles. Higher-order chains exist (e.g. three-member, four-member, five-member chains, etc.) but they are rare.

The pyroxene group consists of 21 mineral species. Pyroxenes have a general structure formula of $XY(Si_2O_6)$, where X is an octahedral site, while Y can vary in coordination number from six to eight. Most varieties of pyroxene consist of permutations of Ca^{2+}, Fe^{2+} and Mg^{2+} to balance the negative charge on the backbone. Pyroxenes are common in the Earth's crust (about 10%) and are a key constituent of mafic igneous rocks.

Amphiboles have great variability in chemistry, described variously as a "mineralogical garbage can" or a "mineralogical shark swimming a sea of elements". The backbone of the amphiboles is the $[Si_8O_{22}]^{12-}$; it is balanced by cations in three possible positions, although the third position is not always used, and one element can occupy both remaining ones. Finally, the amphiboles are usually hydrated, that is, they have a hydroxyl group ($[OH]^-$), although it can be replaced by a fluoride, a chloride, or an oxide ion. Because of the variable chemistry, there are over 80 species of amphibole, although variations, as in the pyroxenes, most commonly involve mixtures of Ca^{2+}, Fe^{2+} and Mg^{2+}. Several amphibole mineral species can have an asbestiform crystal habit. These asbestos minerals form long, thin, flexible, and strong fibres, which are electrical insulators, chemically inert and heat-resistant; as such, they have several applications, especially in construction materials. However, asbestos are known carcinogens, and cause various other illnesses, such as asbestosis; amphibole asbestos (anthophyllite, tremolite, actinolite, grunerite, and riebeckite) are considered more dangerous than chrysotile serpentine asbestos.

Cyclosilicates

Cyclosilicates, or ring silicates, have a ratio of silicon to oxygen of 1:3. Six-member rings are most common, with a base structure of $[Si_6O_{18}]^{12-}$; examples include the tourmaline group and beryl. Other ring structures exist, with 3, 4, 8, 9, 12 having been described. Cyclosilicates tend to be strong, with elongated, striated crystals.

Tourmalines have a very complex chemistry that can be described by a general formula $XY_3Z_6(BO_3)_3T_6O_{18}V_3W$. The T_6O_{18} is the basic ring structure, where T is usually Si^{4+}, but substitutable by Al_{3+} or B^{3+}. Tourmalines can be subgrouped by the occupancy of the X site, and from there further subdivided by the chemistry of the W site. The Y and Z sites can accommodate a variety of cations, especially various transition metals; this variability in structural transition metal content gives the tourmaline group greater variability in colour. Other cyclosilicates include beryl, $Al_2Be_3Si_6O_{18}$, whose varieties include the gemstones emerald (green) and aquamarine (bluish). Cordierite is structurally similar to beryl, and is a common metamorphic mineral.

Sorosilicates

Sorosilicates, also termed disilicates, have tetrahedron-tetrahedron bonding at one oxygen, which results in a 2:7 ratio of silicon to oxygen. The resultant common structural element is the $[Si_2O_7]^{6-}$ group. The most common disilicates by far are members of the epidote group. Epidotes are found in variety of geologic settings, ranging from mid-ocean ridge to granites to metapelites. Epidotes are built around the structure $[(SiO_4)(Si_2O_7)]^{10-}$ structure; for example, the mineral *species* epidote has calcium, aluminium, and ferric iron to charge balance: $Ca_2Al_2(Fe^{3+}, Al)(SiO_4)(Si_2O_7)O(OH)$. The presence of iron as Fe^{3+} and Fe^{2+} helps understand oxygen fugacity, which in turn is a significant factor in petrogenesis.

Other examples of sorosilicates include lawsonite, a metamorphic mineral forming in the blueschist facies (subduction zone setting with low temperature and high pressure), vesuvianite, which takes up a significant amount of calcium in its chemical structure.

Orthosilicates

Orthosilicates consist of isolated tetrahedra that are charge-balanced by other cations. Also termed nesosilicates, this type of silicate has a silicon:oxygen ratio of 1:4 (e.g. SiO_4). Typical orthosilicates tend to form blocky equant crystals, and are fairly hard. Several rock-forming minerals are part of this subclass, such as the aluminosilicates, the olivine group, and the garnet group.

The aluminosilicates—kyanite, andalusite, and sillimanite, all Al_2SiO_5—are structurally composed of one $[SiO_4]^{4-}$ tetrahedron, and one Al^{3+} in octahedral coordination. The remaining Al^{3+} can be in six-fold coordination (kyanite), five-fold (andalusite) or four-fold (sillimanite); which mineral forms in a given environment is depend on pressure and temperature conditions. In the olivine structure, the main olivine series of $(Mg, Fe)_2SiO_4$ consist of magnesium-rich forsterite and iron-rich fayalite. Both iron and magnesium are in octahedral by oxygen. Other mineral species having this structure exist, such as tephroite, Mn_2SiO_4. The garnet group has a general formula of $X_3Y_2(SiO_4)_3$, where X is a large eight-fold coordinated cation, and Y is a smaller six-fold coordinated cation. There are six ideal endmembers of garnet, split into two group. The pyralspite garnets have Al^{3+} in the Y position: pyrope ($Mg_3Al_2(SiO_4)_3$, almandine ($Fe_3Al_2(SiO_4)_3$), and spessartine ($Mn_3Al_2(SiO_4)_3$). The ugrandite garnets have Ca^{2+} in the X position: uvarovite ($Ca_3Cr_2(SiO_4)_3$), grossular ($Ca_3Al_2(SiO_4)_3$) and andradite ($Ca_3Fe_2(SiO_4)_3$). While there are two subgroups of garnet, solid solutions exist between all six end-members.

Other orthosilicates include zircon, staurolite, and topaz. Zircon ($ZrSiO_4$) is useful in geochronology as the Zr^{4+} can be substituted by U^{6+}; furthermore, because of its very resistant structure, it is difficult to reset it as a chronometre. Staurolite is a common metamorphic intermediate-grade index mineral. It has a particularly complicated crystal structure that was only fully described in 1986. Topaz ($Al_2SiO_4(F, OH)_2$, often found in granitic pegmatites associated with tourmaline, is a common gemstone mineral.

Non-silicates

Native Elements: Native elements are those that are not chemically bonded to other elements. This mineral group includes native metals, semi-metals, and non-metals, and various alloys and solid solutions. The metals are held together by metallic bonding, which confers distinctive physical properties such as their shiny metallic lustre, ductility and malleability, and electrical conductivity. Native elements are subdivided into groups by their structure or chemical attributes.

The gold group, with a cubic close-packed structure, includes metals such as gold, silver, and copper. The platinum group is similar in structure to the gold group. The iron-nickel group is characterized by several iron-nickel alloy species. Two examples are kamacite and taenite, which are found in iron meteorites; these species differ by the amount of Ni in the alloy; kamacite has less than 5–7% nickel and is a variety of native iron, whereas the nickel content of taenite ranges from 7–37%. Arsenic group minerals consist of semi-metals, which have only some metallic; for example, they lack the malleability of metals. Native carbon occurs in two allotropes, graphite and diamond; the latter forms at very high pressure in the mantle, which gives it a much stronger structure than graphite.

Sulfides: The sulfide minerals are chemical compounds of one or more metals or semimetals with a sulfur; tellurium, arsenic, or selenium can substitute for the sulfur. Sulfides tend to be soft, brittle minerals with a high specific gravity. Many powdered sulfides, such as pyrite, have a sulfurous smell when powdered. Sulfides are susceptible to weathering, and many readily dissolve in water; these dissolved minerals can be later redeposited, which creates enriched secondary ore deposits. Sulfides are classified by the ratio of the metal or semimetal to the sulfur, such as M:S equal to 2:1, or 1:1. Many sulfide minerals are economically important as metal ores; examples include sphalerite (ZnS), an ore of zinc, galena (PbS), an ore of lead, cinnabar (HgS), an ore of mercury, and molybdenite (MoS_2, an ore of molybdenum. Pyrite (FeS_2), is the most commonly occurring sulfide, and can be found in most geological environments. It is not, however, an ore of iron, but can be instead oxidized to produce sulfuric acid. Related to the sulfides are the rare sulfosalts, in which a metallic element is bonded to sulfur and a semimetal such as antimony, arsenic, or bismuth. Like the sulfides, sulfosalts are typically soft, heavy, and brittle minerals.

Oxides: Oxide minerals are divided into three categories: simple oxides, hydroxides, and multiple oxides. Simple oxides are characterized by O^{2-} as the main anion and primarily ionic bonding. They can be further subdivided by the ratio of oxygen to the cations. The periclase group consists of minerals with a 1:1 ratio. Oxides with a 2:1 ratio include cuprite (Cu_2O) and water ice. Corundum group minerals have a 2:3 ratio, and includes minerals such as corundum (Al_2O_3), and hematite (Fe_2O_3). Rutile group minerals have a ratio of 1:2; the eponymous species, rutile (TiO_2) is the chief ore of titanium; other examples include cassiterite (SnO_2; ore of tin), and pyrolusite (MnO_2; ore of manganese). In hydroxides, the dominant anion is the hydroxyl ion, OH^-. Bauxites are the chief aluminium ore, and are a heterogeneous mixture of the hydroxide minerals diaspore, gibbsite, and bohmite; they form in areas with a very high rate of chemical weathering (mainly tropical conditions).

Finally, multiple oxides are compounds of two metals with oxygen. A major group within this class are the spinels, with a general formula of $X^{2+}Y^{3+}{}_2O_4$. Examples of species include spinel ($MgAl_2O_4$), chromite ($FeCr_2O_4$), and magnetite (Fe_3O_4). The latter is readily distinguishable by its strong magnetism, which occurs as it has iron in two oxidation states ($Fe^{2+}Fe^{3+}{}_2O_4$), which makes it a multiple oxide instead of a single oxide.

Halides: The halide minerals are compounds where a halogen (fluorine, chlorine, iodine, and bromine) is the main anion. These minerals tend to be soft, weak, brittle, and water-soluble. Common examples of halides include halite (NaCl, table salt), sylvite (KCl), fluorite (CaF_2). Halite and sylvite commonly form as evaporites, and can be dominant minerals in chemical sedimentary rocks. Cryolite, Na_3AlF_6, is a key mineral in the extraction of aluminium from bauxites; however, as the only significant occurrence at Ivittuut, Greenland, in a granitic pegmatite, was depleted, synthetic cryolite can be made from fluorite.

Carbonates: The carbonate minerals are those were the main anionic group is carbonate, $[CO_3]^{2-}$. Carbonates tend to be brittle, many have rhombohedral cleavage, and all react with acid. Due to the last characteristic, field geologists often carry dilute hydrochloric acid to distinguish carbonates from non-carbonates. The reaction of acid with carbonates, most commonly found as the polymorph calcite and aragonite ($CaCO_3$), relates to the dissolution and precipitation of the mineral, which is a key in the formation of limestone caves, features within them such as stalactite and stalagmites, and karst landforms. Carbonates are most often formed as biogenic or chemical sediments in marine environments. The carbonate group is structurally a triangle, where a central C^{4+} cation is surrounded by three O^{2-} anions; different groups of minerals form from different arrangements of these triangles. The most common carbonate mineral is calcite, and is the primary constituent of sedimentary limestone and metamorphic marble. Calcite, $CaCO_3$, can have a high magnesium impurity; under high-Mg conditions, its polymorph aragonite will form instead; the marine geochemistry in this regard can be described as an aragonite or calcite sea, depending on which mineral preferentially forms. Dolomite is a double carbonate, with the formula $CaMg(CO_3)_2$. Secondary dolomitization of limestone is common, where calcite or aragonite are converted to dolomite; this reaction increases pore space (the unit cell volume of dolomite is 88% that of calcite), which can create a reservoir for oil and gas. These two minerals species are members of eponymous mineral groups: the calcite group includes carbonates with the general formula XCO_3, and the dolomite group constitutes minerals with general formula $XY(CO_3)_2$.

Sulfates: The sulfate minerals all contain the sulfate anion, $[SO_4]^{2-}$. They tend to be transparent to translucent, soft, and many are fragile. Sulfate minerals commonly form as evaporites, where they precipitate out of evaporating saline waters; alternative, sulfates can also be found in hydrothermal vein systems associated with sulfides, or as oxidation products of sulfides. Sulfates can be subdivided into anhydrous and hydrous minerals. The most common hydrous sulfate by far is gypsum, $CaSO_4$Å"$2H_2O$. It forms as an evaporite, and is associated other evaporites such as calcite and halite; if it incorporates sand grains as it crystallizes, gypsum can form desert roses. Gypsum has very low thermal conductivity and maintains a low temperature when heated as it loses that heat by dehydrating; as such, gypsum is used as an insulator in materials such as plaster and drywall. The anhydrous equivalent of gypsum is anhydrite; it can form directly from seawater in highly arid conditions. The barite group has the general formula XSO_4, where the X is a large 12-coordinated cation. Examples include barite ($BaSO_4$), celestine ($SrSO_4$), and anglesite ($PbSO_4$); anhydrite is not part of the barite group, as the smaller Ca^{2+} is only in eight-fold coordination.

Phosphates: The phosphate minerals are characterized by the tetrahedral $[PO_4]^{3-}$ unit, although the structure can be generalized, and phosphorus is replaced by antimony, arsenic, or vanadium. The most common phosphate is the apatite group; common species within this group are fluorapatite ($Ca_5(PO_4)_3F$), chlorapatite ($Ca_5(PO_4)_3Cl$) and hydroxylapatite ($Ca_5(PO_4)_3(OH)$). Minerals in this group are the main crystalline constituents of teeth and bones in vertebrates. The relatively abundant monazite group has a general structure of ATO_4, where T is phosphorus or arsenic, and A is often a rare-earth element (REE). Monazite is important in two ways: first, as a REE "sink", it can sufficiently concentrate these elements to become an ore; secondly, monazite group elements can incorporate relatively large amounts of uranium and thorium, which can be used to date the rock based on the decay of the U and Th to lead.

Organic Minerals

The Strunz classification includes a class for organic minerals. These rare compounds contain organic carbon, but can be formed by a geologic process. For example, whewellite, $CaC_2O_4.H_2O$ is an oxalate that can be deposited in hydrothermal ore veins. While hydrated calcium oxalate can be found in coal seams and other sedimentary deposits involving organic matter, the hydrothermal occurrence is not considered to be related to biological activity.

Astrobiology

It has been suggested that biominerals could be important indicators of extraterrestrial life and thus could play an important role in the search for past or present life on the planet Mars. Furthermore, organic components (biosignatures) that are often associated with biominerals are believed to play crucial roles in both pre-biotic and biotic reactions.

On January 24, 2014, NASA reported that current studies by the *Curiosity* and *Opportunity* rovers on Mars will now be searching for evidence of ancient life, including a biosphere based on autotrophic, chemotrophic and/or chemolithoautotrophic microorganisms, as well as ancient water, including fluvio-lacustrine environments (plains related to ancient rivers or lakes) that may have been habitable. The search for evidence of habitability, taphonomy (related to fossils), and organic carbon on the planet Mars is now a primary NASA objective.

Ore

An ore is a type of rock that contains sufficient minerals with important elements including metals that can be economically extracted from the rock. The ores are extracted through mining; these are then refined (often via smelting) to extract the valuable element(s).

The grade or concentration of an ore mineral, or metal, as well as its form of occurrence, will directly affect the costs associated with mining the ore. The cost of extraction must thus be weighed against the metal value contained in the rock to determine what ore can be processed and what ore is of too low a grade to be worth mining. Metal ores are generally oxides, sulfides, silicates, or "native" metals (such as native copper) that are not commonly concentrated in the Earth's crust, or "noble" metals (not usually forming compounds) such as gold. The ores must be processed to extract the metals of interest from the waste rock and from the ore minerals. Ore bodies are formed by a variety of geological processes. The process of ore formation is called ore genesis.

Ore Deposits

An ore deposit is an accumulation of ore. Now this is distinct from a mineral resource as defined by the mineral resource classification criteria. An ore deposit is one occurrence of a particular ore type. Most ore deposits are named according to either their location (for example, the Witswatersrand, South Africa), or after a discoverer (e.g. the kambalda nickel shoots are named after drillers), or after some whimsy, a historical figure, a prominent person, something from mythology (phoenix, kraken, serepentleopard, etc.) or the code name of the resource company which found it (e.g. MKD-5 is the in-house name for the Mount Keith nickel).

Classification

Ore deposits are classified according to various criteria developed via the study of economic geology, or ore genesis. The classifications below are typical.

Hydrothermal Epigenetic Deposits

- *Mesothermal* lode gold deposits, typified by the Golden Mile, Kalgoorlie
- Archaean conglomerate hosted gold-uranium deposits, typified by Elliot Lake, Canada and Witwatersrand, South Africa
- Carlin–type gold deposits, including;
- *Epithermal* stockwork vein deposits

Granite Related Hydrothermal

- IOCG or iron oxide copper gold deposits, typified by the supergiant Olympic Dam Cu-Au-U deposit
- Porphyry copper +/- gold +/- molybdenum +/- silver deposits
- Intrusive-related copper-gold +/- (tin-tungsten), typified by the Tombstone, Arizona deposits
- Hydromagmatic magnetite iron ore deposits and skarns
- Skarn ore deposits of copper, lead, zinc, tungsten, etcetera

Magmatic Deposits

- Magmatic nickel-copper-iron-PGE deposits including
    - o Cumulate vanadiferous or platinum-bearing magnetite or chromite
    - o Cumulate hard-rock titanium (ilmenite) deposits
    - o Komatiite hosted Ni-Cu-PGE deposits

    - Subvolcanic feeder subtype, typified by Noril'sk-Talnakh and the Thompson Belt, Canada
    - Intrusive-related Ni-Cu-PGE, typified by Voisey's Bay, Canada and Jinchuan, China
- Lateritic nickel ore deposits, examples include Goro and Acoje, (Philippines) and Ravensthorpe, Western Australia.

Volcanic-related Deposits

- Volcanic hosted massive sulfide (VHMS) Cu-Pb-Zn including;
    - Examples include Teutonic Bore and Golden Grove, Western Australia
        - Besshi type
        - Kuroko type

Metamorphically Reworked Deposits

- Podiform serpentinite-hosted paramagmatic iron oxide-chromite deposits, typified by Savage River, Tasmania iron ore, Coobina chromite deposit
- Broken Hill Type Pb-Zn-Ag, considered to be a class of reworked SEDEX deposits

Carbonatite-alkaline Igneous Related

- Phosphorus-tantalite-vermiculite (Phalaborwa South Africa)
- Rare earth elements - Mount Weld, Australia and Bayan Obo, Mongolia
- Diatreme hosted diamond in kimberlite, lamproite or lamprophyre

Sedimentary Deposits

- Banded iron formation iron ore deposits, including
    - Channel-iron deposits or pisolite type iron ore
- Heavy mineral sands ore deposits and other sand dune hosted deposits
- Alluvial gold, diamond, tin, platinum or black sand deposits
- Alluvial oxide zinc deposit type: sole example Skorpion Zinc

Sedimentary Hydrothermal Deposits

- SEDEX
    - Lead-zinc-silver, typified by Red Dog, McArthur River, Mount Isa, etc.

- o Stratiform arkose-hosted and shale-hosted copper, typified by the Zambian copperbelt.
- o Stratiform tungsten, typified by the Erzgebirge deposits, Czechoslovakia
- o Exhalative spilite-chert hosted gold deposits

- Mississippi valley type (MVT) zinc-lead deposits
- Hematite iron ore deposits of altered banded iron formation

Astrobleme-related Ores

- Sudbury Basin nickel and copper, Ontario, Canada

Extraction

The basic extraction of ore deposits follows these steps:

1. Prospecting or exploration to find and then define the extent and value of ore where it is located ("ore body")
2. Conduct resource estimation to mathematically estimate the size and grade of the deposit
3. Conduct a pre-feasibility study to determine the theoretical economics of the ore deposit. This identifies, early on, whether further investment in estimation and engineering studies is warranted and identifies key risks and areas for further work.
4. Conduct a feasibility study to evaluate the financial viability, technical and financial risks and robustness of the project and make a decision as whether to develop or walk away from a proposed mine project. This includes mine planning to evaluate the economically recoverable portion of the deposit, the metallurgy and ore recoverability, marketability and payability of the ore concentrates, engineering, milling and infrastructure costs, finance and equity requirements and a cradle to grave analysis of the possible mine, from the initial excavation all the way through to reclamation.
5. Development to create access to an ore body and building of mine plant and equipment
6. The operation of the mine in an active sense
7. Reclamation to make land where a mine had been suitable for future use

Trade

Ores (metals) are traded internationally and comprise a sizeable portion of international trade in raw materials both in value and volume. This is because the worldwide distribution of ores is unequal and dislocated from locations of peak demand and from smelting infrastructure. Most base metals (copper, lead, zinc, nickel) are traded internationally on the London Metal Exchange, with smaller stockpiles and metals exchanges monitored by the COMEX and NYMEX exchanges in the United States and the Shanghai Futures Exchange in China.

Iron ore is traded between customer and producer, though various benchmark prices are set quarterly between the major mining conglomerates and the major consumers, and this sets the stage for smaller participants.

Other, lesser, commodities do not have international clearing houses and benchmark prices, with most prices negotiated between suppliers and customers one-on-one. This generally makes determining the price of ores of this nature opaque and difficult. Such metals include lithium, niobium-tantalum, bismuth, antimony and rare earths. Most of these commodities are also dominated by one or two major suppliers with >60% of the world's reserves. The London Metal Exchange aims to add uranium to its list of metals on warrant.

The World Bank reports that China was the top importer of ores and metals in 2005 followed by the USA and Japan.

Metal

A metal (from Greek "μÝταλλïν" – *métallon*, "mine, quarry, metal") is a material (an element, compound, or alloy) that is typically hard, opaque, shiny, and has good electrical and thermal conductivity. Metals are generally malleable — that is, they can be hammered or pressed permanently out of shape without breaking or cracking — as well as fusible (able to be fused or melted) and ductile (able to be drawn out into a thin wire). About 91 of the 118 elements in the periodic table are metals (some elements appear in both metallic and non-metallic forms).

The meaning of "metal" differs for various communities. For example, astronomers use the blanket term "metal" for convenience to collectively describe all elements other than hydrogen and helium (the main components of stars, which in turn comprise most of the visible matter in the universe). Thus, in astronomy and physical cosmology, the metallicity of an object is the proportion of its matter made up of chemical elements other than hydrogen and helium. In addition, many elements and compounds that are not normally classified as metals become metallic under high pressures; these are formed as metallic allotropes of non-metals.

Structure and Bonding

The atoms of metallic substances are closely positioned to neighbouring atoms in one of two common arrangements. The first arrangement is known as body-centred cubic. In this arrangement, each atom is positioned at the centre of eight others. The other is known as face-centred cubic. In this arrangement, each atom is positioned in the centre of six others. The ongoing arrangement of atoms in these structures forms a crystal. Some metals adopt both structures depending on the temperature.

Atoms of metals readily lose their outer shell electrons, resulting in a free flowing cloud of electrons within their otherwise solid arrangement. This provides the ability of metallic substances to easily transmit heat and electricity. While this flow of electrons occurs, the solid characteristic of the metal is produced by electrostatic interactions between each atom and the electron cloud. This type of bond is called a metallic bond.

Properties

Chemical

Metals are usually inclined to form cations through electron loss, reacting with oxygen in the air to form oxides over various timescales (iron rusts over years, while potassium burns in seconds). Examples:

$$4\ Na + O_2 \rightarrow 2\ Na_2O \text{ (sodium oxide)}$$

$$2\ Ca + O_2 \rightarrow 2\ CaO \text{ (calcium oxide)}$$

$$4\ Al + 3\ O_2 \rightarrow 2\ Al_2O_3 \text{ (aluminium oxide).}$$

The transition metals (such as iron, copper, zinc, and nickel) are slower to oxidize because they form a passivating layer of oxide that protects the interior. Others, like palladium, platinum and gold, do not react with the atmosphere at all. Some metals form a barrier layer of oxide on their surface which cannot be penetrated by further oxygen molecules and thus retain their shiny appearance and good conductivity for many decades (like aluminium, magnesium, some steels, and titanium). The oxides of metals are generally basic, as opposed to those of nonmetals, which are acidic. Blatant exceptions are largely oxides with very high oxidation states such as CrO_3, Mn_2O_7, and OsO_4, which have strictly acidic reactions.

Painting, anodizing or plating metals are good ways to prevent their corrosion. However, a more reactive metal in the electrochemical series must be chosen for coating, especially when chipping of the coating is expected. Water and the two metals form an electrochemical cell, and if the coating is less reactive than the coatee, the coating actually *promotes* corrosion.

Physical

Metals in general have high electrical conductivity, high thermal conductivity, and high density. Typically they are malleable and ductile, deforming under stress without cleaving. In terms of optical properties, metals are shiny and lustrous. Sheets of metal beyond a few micrometres in thickness appear opaque, but gold leaf transmits green light.

Although most metals have higher densities than most nonmetals, there is wide variation in their densities, Lithium being the least dense solid element and osmium the densest. The alkali and alkaline earth metals in groups I A and II A are referred to as the light metals because they have low density, low hardness, and low melting points.

The high density of most metals is due to the tightly packed crystal lattice of the metallic structure. The strength of metallic bonds for different metals reaches a maximum around the centre of the transition metal series, as those elements have large amounts of delocalized electrons in tight binding type metallic bonds. However, other factors (such as atomic radius, nuclear charge, number of bonds orbitals, overlap of orbital energies and crystal form) are involved as well.

Electrical

The electrical and thermal conductivities of metals originate from the fact that their outer electrons are delocalized. This situation can be visualized by seeing the atomic structure of a metal as a collection of atoms embedded in a sea of highly mobile electrons. The electrical conductivity, as well as the electrons' contribution to the heat capacity and heat conductivity of metals can be calculated from the free electron model, which does not take into account the detailed structure of the ion lattice.

When considering the electronic band structure and binding energy of a metal, it is necessary to take into account the positive potential caused by the specific arrangement of the ion cores – which is periodic in crystals.

The most important consequence of the periodic potential is the formation of a small band gap at the boundary of the Brillouin zone. Mathematically, the potential of the ion cores can be treated by various models, the simplest being the nearly free electron model.

Mechanical

Mechanical properties of metals include ductility, i.e. their capacity for plastic deformation. Reversible elastic deformation in metals can be described by Hooke's Law for restoring forces, where the stress is linearly proportional to the strain. Forces larger than the elastic limit, or heat, may cause a permanent (irreversible) deformation of the object, known as plastic deformation or plasticity. This irreversible change in atomic arrangement may occur as a result of:

- The action of an applied force (or work). An applied force may be tensile (pulling) force, compressive (pushing) force, shear, bending or torsion (twisting) forces.
- A change in temperature (heat). A temperature change may affect the mobility of the structural defects such as grain boundaries, point vacancies, line and screw dislocations, stacking faults and twins in both crystalline and non-crystalline solids. The movement or displacement of such mobile defects is thermally activated, and thus limited by the rate of atomic diffusion.

Viscous flow near grain boundaries, for example, can give rise to internal slip, creep and fatigue in metals. It can also contribute to significant changes in the microstructure like grain growth and localized densification due to the elimination of intergranular porosity. Screw dislocations may slip in the direction of any lattice plane containing the dislocation, while the principal driving force for "dislocation climb" is the movement or diffusion of vacancies through a crystal lattice.

In addition, the nondirectional nature of metallic bonding is also thought to contribute significantly to the ductility of most metallic solids. When the planes of an ionic bond slide past one another, the resultant change in location shifts ions of the same charge into close proximity, resulting in the cleavage of the crystal; such shift is not observed in covalently bonded crystals where fracture and crystal fragmentation occurs.

Alloys

An alloy is a mixture of two or more elements in which the main component is a metal. Most pure metals are either too soft, brittle or chemically reactive for practical use. Combining different ratios of metals as alloys modifies the properties of pure metals to produce desirable characteristics. The aim of making alloys is generally to make them less brittle, harder, resistant to corrosion, or have a more desirable colour and luster. Of all the metallic alloys in use today, the alloys of iron (steel, stainless steel, cast iron, tool steel, alloy steel) make up the largest proportion both by quantity and commercial value. Iron alloyed with various proportions of carbon gives low, mid and high carbon steels, with increasing carbon levels reducing ductility and toughness. The addition of silicon will produce cast irons, while the addition of chromium, nickel and molybdenum to carbon steels (more than 10%) results in stainless steels.

Other significant metallic alloys are those of aluminium, titanium, copper and magnesium. Copper alloys have been known since prehistory—bronze gave the Bronze Age its name—and have many applications today, most importantly in electrical wiring. The alloys of the other three metals have been developed relatively recently; due to their chemical reactivity they require electrolytic extraction processes. The alloys of aluminium, titanium and magnesium are valued for their high strength-to-weight ratios; magnesium can also provide electromagnetic shielding. These materials are ideal for situations where high strength-to-weight ratio is more important than material cost, such as in aerospace and some automotive applications. Alloys specially designed for highly demanding applications, such as jet engines, may contain more than ten elements.

Categories

Base Metal

In chemistry, the term *base metal* is used informally to refer to a metal that oxidizes or corrodes relatively easily, and reacts variably with dilute hydrochloric acid (HCl) to form hydrogen. Examples include iron, nickel, lead and zinc. Copper is considered a base metal as it oxidizes relatively easily, although it does not react with HCl. It is commonly used in opposition to noble metal.

In alchemy, a *base metal* was a common and inexpensive metal, as opposed to precious metals, mainly gold and silver. A longtime goal of the alchemists was the transmutation of base metals into precious metals.

In numismatics, coins in the past derived their value primarily from the precious metal content. Most modern currencies are fiat currency, allowing the coins to be made of *base metal.*

Ferrous Metal

The term "ferrous" is derived from the Latin word meaning "containing iron". This can include pure iron, such as wrought iron, or an alloy such as steel. Ferrous metals are often magnetic, but not exclusively.

Noble Metal

Noble metals are metals that are resistant to corrosion or oxidation, unlike most base metals. They tend to be precious metals, often due to perceived rarity. Examples include gold, platinum, silver and rhodium.

Precious Metal

A *precious metal* is a rare metallic chemical element of high economic value. Chemically, the precious metals are less reactive than most elements, have high luster and high electrical conductivity. Historically, precious metals were important as currency, but are now regarded mainly as investment and industrial commodities. Gold, silver, platinum and palladium each have an ISO 4217 currency code. The best-known precious metals are gold and silver. While both have industrial uses, they are better known for their uses in art, jewellery, and coinage. Other precious metals include the platinum group metals: ruthenium, rhodium, palladium, osmium, iridium, and platinum, of which platinum is the most widely traded.

The demand for precious metals is driven not only by their practical use, but also by their role as investments and a store of value. Palladium was, as of summer 2006, valued at a little under half the price of gold, and platinum at around twice that of gold. Silver is substantially less expensive than these metals, but is often traditionally considered a precious metal for its role in coinage and jewellery.

Extraction

Metals are often extracted from the Earth by means of mining, resulting in ores that are relatively rich sources of the requisite elements. Ore is located by prospecting techniques, followed by the exploration and examination of deposits. Mineral sources are generally divided into surface mines, which are mined by excavation using heavy equipment, and subsurface mines.

Once the ore is mined, the metals must be extracted, usually by chemical or electrolytic reduction. Pyrometallurgy uses high temperatures to convert ore into raw metals, while hydrometallurgy employs aqueous chemistry for the same purpose. The methods used depend on the metal and their contaminants.

When a metal ore is an ionic compound of that metal and a non-metal, the ore must usually be smelted — heated with a reducing agent — to extract the pure metal. Many common metals, such as iron, are smelted using carbon as a reducing agent. Some metals, such as aluminium and sodium, have no commercially practical reducing agent, and are extracted using electrolysis instead.

Sulfide ores are not reduced directly to the metal but are roasted in air to convert them to oxides.

Recycling of Metals

Demand for metals is closely linked to economic growth. During the 20th century, the variety of metals uses in society grew rapidly. Today, the development of major nations, such as China and India, and advances in technologies, are fuelling ever more demand. The result is that mining activities are expanding, and more and more of the world's metal stocks are above ground in use, rather than below ground as unused reserves. An example is the in-use stock of copper. Between 1932 and 1999, copper in use in the USA rose from 73g to 238g per person.

Metals are inherently recyclable, so in principle, can be used over and over again, minimizing these negative environmental impacts and saving energy at the same time. For example, 95% of the energy used to make aluminium from bauxite ore is saved by using recycled material. However, levels of metals recycling are generally low. In 2010, the International Resource Panel, hosted by the United Nations Environment Programme (UNEP) published reports on metal stocks that exist within society and their recycling rates.

The report authors observed that the metal stocks in society can serve as huge mines above ground. However, they warned that the recycling rates of some rare metals used in applications such as mobile phones, battery packs for hybrid cars and fuel cells are so low that unless future end-of-life recycling rates are dramatically stepped up these critical metals will become unavailable for use in modern technology.

Metallurgy

Metallurgy is a domain of materials science that studies the physical and chemical behaviour of metallic elements, their intermetallic compounds, and their mixtures, which are called alloys.

Applications

Some metals and metal alloys possess high structural strength per unit mass, making them useful materials for carrying large loads or resisting impact damage. Metal alloys can be engineered to have high resistance to shear, torque and deformation. However the same metal can also be vulnerable to fatigue damage through repeated use or from sudden stress failure when a load capacity is exceeded. The strength and resilience of metals has led to their frequent use in high-rise building and bridge construction, as well as most vehicles, many appliances, tools, pipes, non-illuminated signs and railroad tracks.

The two most commonly used structural metals, iron and aluminium, are also the most abundant metals in the Earth's crust.

Metals are good conductors, making them valuable in electrical appliances and for carrying an electric current over a distance with little energy lost. Electrical power grids rely on metal cables to distribute electricity. Home electrical systems, for the most part, are wired with copper wire for its good conducting properties.

The thermal conductivity of metal is useful for containers to heat materials over a flame. Metal is also used for heat sinks to protect sensitive equipment from overheating.

The high reflectivity of some metals is important in the construction of mirrors, including precision astronomical instruments. This last property can also make metallic jewellery aesthetically appealing.

Some metals have specialized uses; radioactive metals such as uranium and plutonium are used in nuclear power plants to produce energy via nuclear fission. Mercury is a liquid at room temperature and is used in switches to complete a circuit when it flows over the switch contacts. Shape memory alloy is used for applications such as pipes, fasteners and vascular stents.

Trade

The World Bank reports that China was the top importer of ores and metals in 2005 followed by the United States and Japan.

Fossil Fuel

Fossil fuels are fuels formed by natural processes such as anaerobic decomposition of buried dead organisms. The age of the organisms and their resulting fossil fuels is typically millions of years, and sometimes exceeds 650 million years. Fossil fuels contain high percentages of carbon and include coal, petroleum, and natural gas. They range from volatile materials with low carbon:hydrogen ratios like methane, to liquid petroleum to nonvolatile materials composed of almost pure carbon, like anthracite coal. Methane can be found in hydrocarbon fields, alone, associated with oil, or in the form of methane clathrates. The theory that fossil fuels formed from the fossilized remains of dead plants by exposure to heat and pressure in the Earth's crust over millions of years was first introduced by Georg Agricola in 1556 and later by Mikhail Lomonosov in the 18th century.

The Energy Information Administration estimates that in 2007 the primary sources of energy consisted of petroleum 36.0%, coal 27.4%, natural gas 23.0%, amounting to an 86.4% share for fossil fuels in primary energy consumption in the world. Non-fossil sources in 2006 included hydroelectric 6.3%, nuclear 8.5%, and others (geothermal, solar, tidal, wind, wood, waste) amounting to 0.9%. World energy consumption was growing about 2.3% per year.

Strictly speaking, fossil fuels are a renewable resource. They are continually being formed via natural processes as plants and animals die and then decompose and become trapped beneath sediment. However, fossil fuels are generally considered to be non-renewable resources because they take millions of years to form, and known viable reserves are being depleted much faster than new ones are being made.

The use of fossil fuels raises serious environmental concerns. The burning of fossil fuels produces around 21.3 billion tonnes (21.3 gigatonnes) of carbon dioxide (CO_2) per year, but it is estimated that natural processes can only absorb about half of that amount, so there is a net increase of 10.65 billion tonnes of atmospheric carbon dioxide per year (one tonne of atmospheric carbon is equivalent to 44/12 or 3.7 tonnes of carbon dioxide). Carbon dioxide is one of the greenhouse gases that enhances radiative forcing and contributes to global warming, causing the average surface temperature of the Earth to rise in response, which the vast majority of climate scientists agree will cause major adverse effects. A global movement towards the generation of renewable energy is therefore under way to help reduce global greenhouse gas emissions.

Origin

Petroleum and natural gas are formed by the anaerobic decomposition of remains of organisms including phytoplankton and zooplankton that settled to the sea (or lake) bottom in large quantities under anoxic conditions, millions of years ago. Over geological time, this organic matter, mixed with mud, got buried under heavy layers of sediment. The resulting high levels of heat and pressure caused the organic matter to chemically alter, first into a waxy material known as kerogen which is found in oil shales, and then with more heat into liquid and gaseous hydrocarbons in a process known as catagenesis.

There is a wide range of organic, or hydrocarbon, compounds in any given fuel mixture. The specific mixture of hydrocarbons gives a fuel its characteristic properties, such as boiling point, melting point, density, viscosity, etc. Some fuels like natural gas, for instance, contain only very low boiling, gaseous components. Others such as gasoline or diesel contain much higher boiling components.

Terrestrial plants, on the other hand, tend to form coal and methane. Many of the coal fields date to the Carboniferous period of Earth's history. Terrestrial plants also form type III kerogen, a source of natural gas.

Importance

Fossil fuels are of great importance because they can be burned (oxidized to carbon dioxide and water), producing significant amounts of energy per unit weight.

The use of coal as a fuel predates recorded history. Coal was used to run furnaces for the melting of metal ore. Semi-solid hydrocarbons from seeps were also burned in ancient times, but these materials were mostly used for waterproofing and embalming.

Commercial exploitation of petroleum, largely as a replacement for oils from animal sources (notably whale oil), for use in oil lamps began in the 19th century.

Natural gas, once flared-off as an unneeded byproduct of petroleum production, is now considered a very valuable resource. Natural gas deposits are also the main source of the element helium.

Heavy crude oil, which is much more viscous than conventional crude oil, and tar sands, where bitumen is found mixed with sand and clay, are becoming more important as sources of fossil fuel.

Oil shale and similar materials are sedimentary rocks containing kerogen, a complex mixture of high-molecular weight organic compounds, which yield synthetic crude oil when heated (pyrolyzed). These materials have yet to be exploited commercially. These fuels can be employed in internal combustion engines, fossil fuel power stations and other uses.

Prior to the latter half of the 18th century, windmills and watermills provided the energy needed for industry such as milling flour, sawing wood or pumping water, and burning wood or peat provided domestic heat. The widescale use of fossil fuels, coal at first and petroleum later, to fire steam engines enabled the Industrial Revolution. At the same time, gas lights using natural gas or coal gas were coming into wide use.

The invention of the internal combustion engine and its use in automobiles and trucks greatly increased the demand for gasoline and diesel oil, both made from fossil fuels. Other forms of transportation, railways and aircraft, also required fossil fuels. The other major use for fossil fuels is in generating electricity and as feedstock for the petrochemical industry. Tar, a leftover of petroleum extraction, is used in construction of roads.

Reserves

Levels of primary energy sources are the reserves in the ground. Flows are production of fossil fuels from these reserves. The most important part of primary energy sources are the carbon based fossil energy sources.

Coal, oil, and natural gas provided 79.6% of primary energy production during 2002 (in million tonnes of oil equivalent (mtoe)) (34.9+23.5+21.2).

Levels (proved reserves) during 2005–2006

- Coal: 997,748 million short tonnes (905 billion metric tonnes), 4,416 billion barrels (702.1 km^3) of oil equivalent
- Oil: 1,119 billion barrels (177.9 km^3) to 1,317 billion barrels (209.4 km^3)
- Natural gas: 6,183–6,381 trillion cubic feet (175–181 trillion cubic metres), 1,161 billion barrels (184.6×10^9 m^3) of oil equivalent

Flows (daily production) during 2006

- Coal: 18,476,127 short tonnes (16,761,260 metric tonnes), 52,000,000 barrels (8,300,000 m^3) of oil equivalent per day
- Oil: 84,000,000 barrels per day (13,400,000 m^3/d)
- Natural gas: 104,435 billion cubic feet (2,963 billion cubic metres), 19,000,000 barrels (3,000,000 m^3) of oil equivalent per day

Limits and Alternatives

P. E. Hodgson, a Senior Research Fellow Emeritus in Physics at Corpus Christi College, Oxford, expects the world energy use is doubling every fourteen years and the need is increasing faster still and he insisted in 2008 that the world oil production, a main resource of fossil fuel, is expected to peak in ten years and thereafter fall.

The principle of supply and demand holds that as hydrocarbon supplies diminish, prices will rise. Therefore higher prices will lead to increased alternative, renewable energy supplies as previously uneconomic sources become sufficiently economical to exploit.

Artificial gasolines and other renewable energy sources currently require more expensive production and processing technologies than conventional petroleum reserves, but may become economically viable in the near future. Different alternative sources of energy include nuclear, hydroelectric, solar, wind, and geothermal.

Environmental Effects

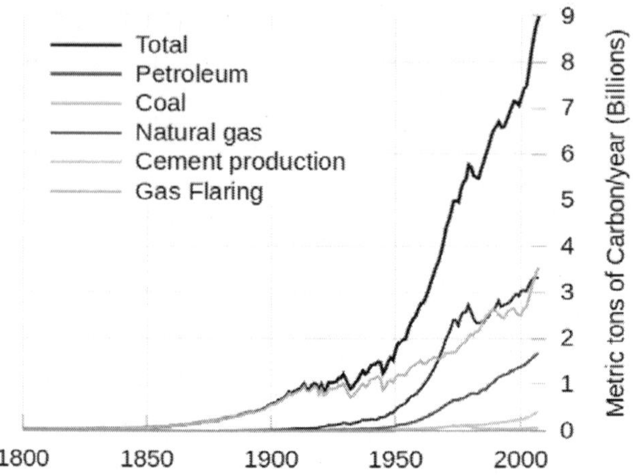

Figure: *Global fossil carbon emission by fuel type, 1800–2007. Note: Carbon only represents 27% of the mass of CO2*

The U.S. holds less than 5% of the world's population, but due to large houses and private cars, uses more than a quarter of the world's supply of fossil fuels. In the United States, more than 90% of greenhouse gas emissions come from the combustion of fossil fuels. Combustion of fossil fuels also produces other air pollutants, such as nitrogen oxides, sulfur dioxide, volatile organic compounds and heavy metals.

According to Environment Canada

"The electricity sector is unique among industrial sectors in its very large contribution to emissions associated with nearly all air issues. Electricity generation produces a large share of Canadian nitrogen oxides and sulphur dioxide emissions, which contribute to smog and acid rain and the formation of fine particulate matter. It is the largest uncontrolled industrial source of mercury emissions in Canada. Fossil fuel-fired electric power plants also emit carbon dioxide, which may contribute to climate change. In addition, the sector has significant impacts on water and habitat and species. In particular, hydro dams and transmission lines have significant effects on water and biodiversity."

According to U.S. Scientist Jerry Mahlman and USA Today: Mahlman, who crafted the IPCC language used to define levels of scientific certainty, says the new report will lay the blame at the feet of fossil fuels with "virtual certainty," meaning 99% sure. That's a significant jump from "likely," or 66% sure, in the group's last report in 2001, Mahlman says. His role in this year's effort involved spending two months reviewing the more than 1,600 pages of research that went into the new assessment.

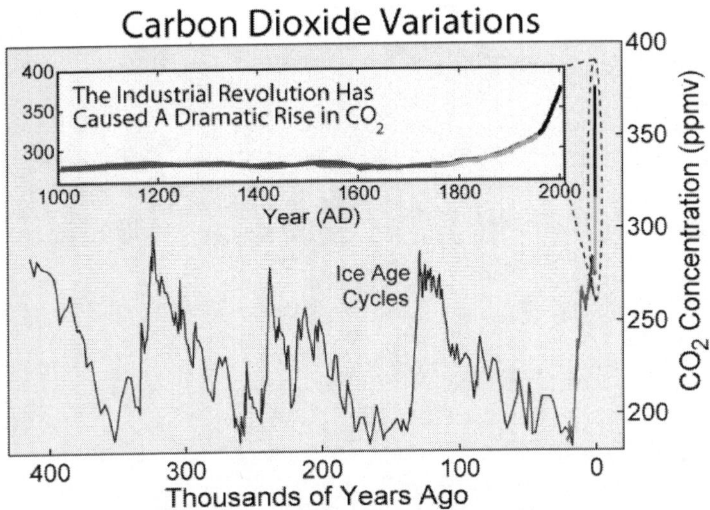

Figure: *Carbon dioxide variations over the last 400,000 years, showing a rise since the industrial revolution.*

Combustion of fossil fuels generates sulfuric, carbonic, and nitric acids, which fall to Earth as acid rain, impacting both natural areas and the built environment. Monuments and sculptures made from marble and limestone are particularly vulnerable, as the acids dissolve calcium carbonate.

Fossil fuels also contain radioactive materials, mainly uranium and thorium, which are released into the atmosphere. In 2000, about 12,000 tonnes of thorium and 5,000 tonnes of uranium were released worldwide from burning coal. It is estimated that during 1982, US coal burning released 155 times as much radioactivity into the atmosphere as the Three Mile Island incident.

Burning coal also generates large amounts of bottom ash and fly ash. These materials are used in a wide variety of applications, utilizing, for example, about 40% of the US production.

Harvesting, processing, and distributing fossil fuels can also create environmental concerns. Coal mining methods, particularly mountaintop removal and strip mining, have negative environmental impacts, and offshore oil drilling poses a hazard to aquatic organisms. Oil refineries also have negative environmental impacts, including air and water pollution. Transportation of coal requires the use of diesel-powered locomotives, while crude oil is typically transported by tanker ships, each of which requires the combustion of additional fossil fuels.

Environmental regulation uses a variety of approaches to limit these emissions, such as command-and-control (which mandates the amount of pollution or the technology used), economic incentives, or voluntary programs. An example of such regulation in the USA is the "EPA is implementing policies to reduce airborne mercury emissions. Under regulations issued in 2005, coal-fired power plants will need to reduce their emissions by 70 percent by 2018.".

In economic terms, pollution from fossil fuels is regarded as a negative externality. Taxation is considered one way to make societal costs explicit, in order to 'internalize' the cost of pollution. This aims to make fossil fuels more expensive, thereby reducing their use and the amount of pollution associated with them, along with raising the funds necessary to counteract these factors.

According to Rodman D. Griffin, "The burning of coal and oil have saved inestimable amounts of time and labour while substantially raising living standards around the world". Although the use of fossil fuels may seem beneficial to our lives, this act is playing a role on global warming and it is said to be dangerous for the future.

Moreover, these environmental pollutions impacts on the human beings because its particles of the fossil fuel on the air cause negative health effects when inhaled by people. These health effects include premature death, acute respiratory illness, aggravated asthma, chronic bronchitis and decreased lung function. So, the poor, undernourished, very young and very old, and people with preexisting respiratory disease and other ill health, are more at risk.

Economical Effects

Europe spent on importing fossil fuels €406 billion in 2011 (bn) and €545 bn in 2012. This is around three times more than the cost of the Greek bailout up to 2013. In 2012 wind energy in Europe avoided €9.6 billion (bn) of fossil fuel costs.

Chapter 4

Renewable Energy

Renewable energy is generally defined as energy that comes from resources which are naturally replenished on a human timescale such as sunlight, wind, rain, tides, waves and geothermal heat. Renewable energy replaces conventional fuels in four distinct areas: electricity generation, hot water/space heating, motor fuels, and rural (off-grid) energy services.

Based on REN21's 2014 report, renewables contributed 19 percent to our energy consumption and 22 percent to our electricity generation in 2012 and 2013, respectively. Both, modern renewables, such as hydro, wind, solar and biofuels, as well as traditional biomass, contributed in about equal parts to the global energy supply. Worldwide investments in renewable technologies amounted to more than US$ 214 billion in 2013, with countries like China and the United States heavily investing in wind, hydro, solar and biofuels.

Renewable energy resources exist over wide geographical areas, in contrast to other energy sources, which are concentrated in a limited number of countries. Rapid deployment of renewable energy and energy efficiency is resulting in significant energy security, climate change mitigation, and economic benefits. In international public opinion surveys there is strong support for promoting renewable sources such as solar power and wind power. At the national level, at least 30 nations around the world already have renewable energy contributing more than 20 percent of energy supply. National renewable energy markets are projected to continue to grow strongly in the coming decade and beyond.

While many renewable energy projects are large-scale, renewable technologies are also suited to rural and remote areas and developing countries, where energy is often crucial in human development. United Nations' Secretary-General Ban Ki-moon has said that renewable energy has the ability to lift the poorest nations to new levels of prosperity.

Overview

Renewable energy flows involve natural phenomena such as sunlight, wind, tides, plant growth, and geothermal heat, as the International Energy Agency explains:

Renewable energy is derived from natural processes that are replenished constantly. In its various forms, it derives directly from the sun, or from heat generated deep within the earth. Included in the definition is electricity and heat generated from solar, wind, ocean, hydropower, biomass, geothermal resources, and biofuels and hydrogen derived from renewable resources.

Wind power is growing at the rate of 30% annually, with a worldwide installed capacity of 282,482 megawatts (MW) at the end of 2012, and is widely used in Europe, Asia, and the United States. At the end of 2012 the photovoltaic (PV) capacity worldwide was 100,000 MW, and PV power stations are popular in Germany and Italy. Solar thermal power stations operate in the USA and Spain, and the largest of these is the 354 MW SEGS power plant in the Mojave Desert. The world's largest geothermal power installation is The Geysers in California, with a rated capacity of 750 MW. Brazil has one of the largest renewable energy programs in the world, involving production of ethanol fuel from sugar cane, and ethanol now provides 18% of the country's automotive fuel. Ethanol fuel is also widely available in the USA.

Renewable energy resources and significant opportunities for energy efficiency exist over wide geographical areas, in contrast to other energy sources, which are concentrated in a limited number of countries. Rapid deployment of renewable energy and energy efficiency, and technological diversification of energy sources, would result in significant energy security and economic benefits.

Renewable energy replaces conventional fuels in four distinct areas: electricity generation, hot water/space heating, motor fuels, and rural (off-grid) energy services:

- Power generation. Renewable energy provides 21.7% of electricity generation worldwide as of 2013. Renewable power generators are spread across many countries, and wind power alone already provides a significant share of electricity in some areas: for example, 14% in the U.S. state of Iowa, 40% in the northern German state of Schleswig-Holstein, and 49% in

Denmark. Some countries get most of their power from renewables, including Iceland (100%), Norway (98%), Brazil (86%), Austria (62%), New Zealand (65%), and Sweden (54%).

- Heating. Solar hot water makes an important contribution to renewable heat in many countries, most notably in China, which now has 70% of the global total (180 GWth). Most of these systems are installed on multi-family apartment buildings and meet a portion of the hot water needs of an estimated 50–60 million households in China. Worldwide, total installed solar water heating systems meet a portion of the water heating needs of over 70 million households. The use of biomass for heating continues to grow as well. In Sweden, national use of biomass energy has surpassed that of oil. Direct geothermal for heating is also growing rapidly.
- Transport fuels. Renewable biofuels have contributed to a significant decline in oil consumption in the United States since 2006. The 93 billion liters of biofuels produced worldwide in 2009 displaced the equivalent of an estimated 68 billion liters of gasoline, equal to about 5% of world gasoline production.

As of 2011, small solar PV systems provide electricity to a few million households, and micro-hydro configured into mini-grids serves many more. Over 44 million households use biogas made in household-scale digesters for lighting and/or cooking, and more than 166 million households rely on a new generation of more-efficient biomass cookstoves. United Nations' Secretary-General Ban Ki-moon has said that renewable energy has the ability to lift the poorest nations to new levels of prosperity.

At the national level, at least 30 nations around the world already have renewable energy contributing more than 20% of energy supply. National renewable energy markets are projected to continue to grow strongly in the coming decade and beyond, and some 120 countries have various policy targets for longer-term shares of renewable energy, including a 20% target of all electricity generated for the European Union by 2020. Some countries have much higher long-term policy targets of up to 100% renewables. Outside Europe, a diverse group of 20 or more other countries target renewable energy shares in the 2020–2030 time frame that range from 10% to 50%.

Climate change and global warming concerns, coupled with high oil prices, peak oil, and increasing government support, are driving increasing renewable energy legislation, incentives and commercialization. New government spending, regulation and policies helped the industry weather the global financial crisis better than many other sectors. According to a 2011 projection by the International Energy Agency, solar power generators may produce most of the world's electricity within 50 years, reducing the emissions of greenhouse gases that harm the environment.

Renewable energy sources, that derive their energy from the sun, either directly or indirectly, such as hydro and wind, are expected to be capable of supplying humanity energy for almost another 1 billion years, at which point the predicted increase in heat from the sun is expected to make the surface of the earth too hot for liquid water to exist.

History

Prior to the development of coal in the mid 19th century, nearly all energy used was renewable. Almost without a doubt the oldest known use of renewable energy, in the form of traditional biomass to fuel fires, dates from 790,000 years ago. Use of biomass for fire did not become commonplace until many hundreds of thousands of years later, sometime between 200,000 and 400,000 years ago.

Probably the second oldest usage of renewable energy is harnessing the wind in order to drive ships over water. This practice can be traced back some 7000 years, to ships on the Nile.

Moving into the time of recorded history, the primary sources of traditional renewable energy were human labour, animal power, water power, wind, in grain crushing windmills, and firewood, a traditional biomass. A graph of energy use in the United States up until 1900 shows oil and natural gas with about the same importance in 1900 as wind and solar played in 2010.

By 1873, concerns of running out of coal prompted experiments with using solar energy. Development of solar engines continued until the outbreak of World War I. The importance of solar energy was recognized in a 1911 *Scientific American* article: "in the far distant future, natural fuels having been exhausted [solar power] will remain as the only means of existence of the human race".

The theory of peak oil was published in 1956. In the 1970s environmentalists promoted the development of renewable energy both as a replacement for the eventual depletion of oil, as well as for an escape from dependence on oil, and the first electricity generating wind turbines appeared. Solar had long been used for heating and cooling, but solar panels were too costly to build solar farms until 1980.

The IEA 2014 World Energy Outlook projects a growth of renewable energy supply from 1700 gigawatts in 2014 to 4550 gigawatts in 2040. Fossil fuels received about $550 billion in subsidies in 2013, compared to $120 billion for all renewable energies.

Mainstream Technologies

Wind Power

Airflows can be used to run wind turbines. Modern utility-scale wind turbines range from around 600 kW to 5 MW of rated power, although turbines with rated output of 1.5–3 MW have become the most common for commercial use; the power available from the wind is a function of the cube of the wind speed, so as wind speed increases, power output increases up to the maximum output for the particular turbine. Areas where winds are stronger and more constant, such as offshore and high altitude sites, are preferred locations for wind farms. Typical capacity factors are 20-40%, with values at the upper end of the range in particularly favourable sites.

Globally, the long-term technical potential of wind energy is believed to be five times total current global energy production, or 40 times current electricity demand, assuming all practical barriers needed were overcome. This would require wind turbines to be installed over large areas, particularly in areas of higher wind resources, such as offshore. As offshore wind speeds average ~90% greater than that of land, so offshore resources can contribute substantially more energy than land stationed turbines.

Hydropower

Energy in water can be harnessed and used. Since water is about 800 times denser than air, even a slow flowing stream of water, or moderate sea swell, can yield considerable amounts of energy. There are many forms of water energy:

- Hydroelectric energy is a term usually reserved for large-scale hydroelectric dams. The largest of which is the Three Gorges Dam in China and a smaller example is the Akosombo Dam in Ghana.
- Micro hydro systems are hydroelectric power installations that typically produce up to 100 kW of power. They are often used in water rich areas as a remote-area power supply (RAPS).
- Run-of-the-river hydroelectricity systems derive kinetic energy from rivers and oceans without the creation of a large reservoir.

Hydropower is produced in 150 countries, with the Asia-Pacific region generating 32 percent of global hydropower in 2010. China is the largest hydroelectricity producer, with 721 terawatt-hours of production in 2010, representing around 17 percent of domestic electricity use. There are now three hydroelectricity stations larger than 10 GW: the Three Gorges Dam in China, Itaipu Dam across the Brazil/Paraguay border, and Guri Dam in Venezuela.

Wave power, that captures the energy of ocean surface waves, and tidal power, converting the energy of tides, are two forms of hydropower with future potential, however, not yet widely employed commercially, while ocean thermal energy conversion, that uses the temperature difference between cooler deep and warmer surface waters, has currently no economic feasibility.

Solar Energy

Solar energy, radiant light and heat from the sun, is harnessed using a range of ever-evolving technologies such as solar heating, photovoltaics, concentrated solar power, solar architecture and artificial photosynthesis. Solar technologies are broadly characterized as either passive solar or active solar depending on the way they capture, convert and distribute solar energy. Passive solar techniques include orienting a building to the Sun, selecting materials with favourable thermal mass or light dispersing properties, and designing spaces that naturally circulate air.

Active solar technologies encompass solar thermal energy, using solar collectors for heating, and solar power, converting sunlight into electricity either directly using photovoltaics (PV), or indirectly using concentrated solar power (CSP).

A photovoltaic system converts light into electrical direct current (DC) by taking advantage of the photoelectric effect. Solar PV has turned into a multi-billion, fast-growing industry, continues to improve its cost-effectiveness, and has the most potential of any renewable technology. Concentrated solar power systems use lenses or mirrors and tracking systems to focus a large area of sunlight into a small beam. Commercial concentrated solar power plants were first developed in the 1980s. In 2011, the International Energy Agency said that "the development of affordable, inexhaustible and clean solar energy technologies will have huge longer-term benefits. It will increase countries' energy security through reliance on an indigenous, inexhaustible and mostly import-independent resource, enhance sustainability, reduce pollution, lower the costs of mitigating climate change, and keep fossil fuel prices lower than otherwise. These advantages are global. Hence the additional costs of the incentives for early deployment should be considered learning investments; they must be wisely spent and need to be widely shared".

Biomass

Biomass is biological material derived from living, or recently living organisms. It most often refers to plants or plant-derived materials which are specifically called lignocellulosic biomass. As an energy source, biomass can either be used directly via combustion to produce heat, or indirectly after converting it to various forms of biofuel. Conversion of biomass to biofuel can be achieved by different methods which are broadly classified into: *thermal*, *chemical*, and *biochemical* methods.

Wood remains the largest biomass energy source today; examples include forest residues (such as dead trees, branches and tree stumps), yard clippings, wood chips and even municipal solid waste. In the second sense, biomass includes plant or animal matter that can be converted into fibres or other industrial chemicals, including biofuels. Industrial biomass can be grown from numerous types of plants, including miscanthus, switchgrass, hemp, corn, poplar, willow, sorghum, sugarcane, bamboo, and a variety of tree species, ranging from eucalyptus to oil palm (palm oil).

Plant energy is produced by crops specifically grown for use as fuel that offer high biomass output per hectare with low input energy. Some examples of these plants are wheat, which typically yield 7.5–8 tonnes of grain per hectare, and straw, which typically yield 3.5–5 tonnes per hectare in the UK. The grain can be used for liquid transportation fuels while the straw can be burned to produce heat or electricity. Plant biomass can also be degraded from cellulose to glucose through a series of chemical treatments, and the resulting sugar can then be used as a first generation biofuel.

Biomass can be converted to other usable forms of energy like methane gas or transportation fuels like ethanol and biodiesel. Rotting garbage, and agricultural and human waste, all release methane gas – also called "landfill gas" or "biogas". Crops, such as corn and sugar cane, can be fermented to produce the transportation fuel, ethanol. Biodiesel, another transportation fuel, can be produced from left-over food products like vegetable oils and animal fats. Also, biomass to liquids (BTLs) and cellulosic ethanol are still under research.

There is a great deal of research involving algal, or algae-derived, biomass due to the fact that it's a non-food resource and can be produced at rates 5 to 10 times those of other types of land-based agriculture, such as corn and soy. Once harvested, it can be fermented to produce biofuels such as ethanol, butanol, and methane, as well as biodiesel and hydrogen.

The biomass used for electricity generation varies by region. Forest by-products, such as wood residues, are common in the United States. Agricultural waste is common in Mauritius (sugar cane residue) and Southeast Asia (rice husks). Animal husbandry residues, such as poultry litter, are common in the UK.

Biofuel

Biofuels include a wide range of fuels which are derived from biomass. The term covers solid biofuels, liquid biofuels, and gaseous biofuels. Liquid biofuels include bioalcohols, such as bioethanol, and oils, such as biodiesel. Gaseous biofuels include biogas, landfill gas and synthetic gas.

Bioethanol is an alcohol made by fermenting the sugar components of plant materials and it is made mostly from sugar and starch crops. These include maize, sugar cane and, more recently, sweet sorghum. The latter crop is particularly suitable for growing in dryland conditions, and is being investigated by ICRISAT for its potential to provide fuel, along with food and animal feed, in arid parts of Asia and Africa.

With advanced technology being developed, cellulosic biomass, such as trees and grasses, are also used as feedstocks for ethanol production. Ethanol can be used as a fuel for vehicles in its pure form, but it is usually used as a gasoline additive to increase octane and improve vehicle emissions. Bioethanol is widely used in the USA and in Brazil. The energy costs for producing bio-ethanol are almost equal to, the energy yields from bio-ethanol. However, according to the European Environment Agency, biofuels do not address global warming concerns.

Biodiesel is made from vegetable oils, animal fats or recycled greases. Biodiesel can be used as a fuel for vehicles in its pure form, but it is usually used as a diesel additive to reduce levels of particulates, carbon monoxide, and hydrocarbons from diesel-powered vehicles. Biodiesel is produced from oils or fats using transesterification and is the most common biofuel in Europe.

Biofuels provided 2.7% of the world's transport fuel in 2010.

Geothermal Energy

Geothermal energy is from thermal energy generated and stored in the Earth. Thermal energy is the energy that determines the temperature of matter. Earth's geothermal energy originates from the original formation of the planet (20%) and from radioactive decay of minerals (80%). The geothermal gradient, which is the difference in temperature between the core of the planet and its surface, drives a continuous conduction of thermal energy in the form of heat from the core to the surface. The adjective *geothermal* originates from the Greek roots *geo*, meaning earth, and *thermos*, meaning heat.

The heat that is used for geothermal energy can be from deep within the Earth, all the way down to Earth's core – 4,000 miles (6,400 km) down. At the core, temperatures may reach over 9,000 °F (5,000 °C). Heat conducts from the core to surrounding rock. Extremely high temperature and pressure cause some rock to melt, which is commonly known as magma. Magma convects upward since it is lighter than the solid rock. This magma then heats rock and water in the crust, sometimes up to 700 °F (371 °C).

From hot springs, geothermal energy has been used for bathing since Paleolithic times and for space heating since ancient Roman times, but it is now better known for electricity generation.

Commercialization

Growth of Renewables

From the end of 2004, worldwide renewable energy capacity grew at rates of 10–60% annually for many technologies. For wind power and many other renewable technologies, growth accelerated in 2009 relative to the previous four years. More wind power capacity was added during 2009 than any other renewable technology. However, grid-connected PV increased the fastest of all renewables technologies, with a 60% annual average growth rate. In 2010, renewable power constituted about a third of the newly built power generation capacities.

Projections vary, but scientists have advanced a plan to power 100% of the world's energy with wind, hydroelectric, and solar power by the year 2030.

According to a 2011 projection by the International Energy Agency, solar power generators may produce most of the world's electricity within 50 years, reducing the emissions of greenhouse gases that harm the environment. Cedric Philibert, senior analyst in the renewable energy division at the IEA said: "Photovoltaic and solar-thermal plants may meet most of the world's demand for electricity by 2060 – and half of all energy needs – with wind, hydropower and biomass plants supplying much of the remaining generation". "Photovoltaic and concentrated solar power together can become the major source of electricity", Philibert said.

Selected renewable energy global indicators	*2008*	*2009*	*2010*	*2011*	*2012*	*2013*
Investment in new renewable capacity (annual) (10^9 USD)	130	160	211	257	244	214
Renewables power capacity (existing) (GWe)	1,140	1,230	1,320	1,360	1,470	1,560
Hydropower capacity (existing) (GWe)	885	915	945	970	990	1,000
Wind power capacity (existing) (GWe)	121	159	198	238	283	318
Solar PV capacity (grid-connected) (GWe)	16	23	40	70	100	139
Solar hot water capacity (existing) (GWth)	130	160	185	232	255	326
Ethanol production (annual) (10^9 litres)	67	76	86	86	83	87
Biodiesel production (annual) (10^9 litres)	12	17.8	18.5	21.4	22.5	26
Countries with policy targets for renewable energy use	79	89	98	118	138	144

Economic Trends

Renewable energy technologies are getting cheaper, through technological change and through the benefits of mass production and market competition. A 2011 IEA report said: "A portfolio of renewable energy technologies is becoming cost-competitive in an increasingly broad range of circumstances, in some cases providing investment opportunities without the need for specific economic support," and added that "cost reductions in critical technologies, such as wind and solar, are set to continue."

Hydro-electricity and geothermal electricity produced at favourable sites are now the cheapest way to generate electricity. Renewable energy costs continue to drop, and the levelised cost of electricity (LCOE) is declining for wind power, solar photovoltaic (PV), concentrated solar power (CSP) and some biomass technologies.

Renewable energy is also the most economic solution for new grid-connected capacity in areas with good resources. As the cost of renewable power falls, the scope of economically viable applications increases. Renewable technologies are now often the most economic solution for new generating capacity. Where "oil-fired generation is the predominant power generation source (e.g. on islands, off-grid and in some countries) a lower-cost renewable solution almost always exists today".

A series of studies by the US National Renewable Energy Laboratory modelled the "grid in the Western US under a number of different scenarios where intermittent renewables accounted for 33 percent of the total power." In the models, inefficiencies in cycling the fossil fuel plants to compensate for the variation in solar and wind energy resulted in an additional cost of "between \$0.47 and \$1.28 to each MegaWatt hour generated"; however, the savings in the cost of the fuels saved "adds up to \$7 billion, meaning the added costs are, at most, two percent of the savings."

Hydroelectricity

The Three Gorges Dam in Hubei, China, has the world's largest instantaneous generating capacity (22,500 MW), with the Itaipu Dam in Brazil/Paraguay in second place (14,000 MW). The Three Gorges Dam is operated jointly with the much smaller Gezhouba Dam (3,115 MW). As of 2012, the total generating capacity of this two-dam complex is 25,615 MW. In 2008, this complex generated 98 TWh of electricity (81 TWh from the Three Gorges Dam and 17 TWh from the Gezhouba Dam), which is 3% more power in one year than the 95 TWh generated by Itaipu in 2008.

Wind Power Development

Wind power is growing at over 20% annually, with a worldwide installed capacity of 238,000 MW at the end of 2011, and is widely used in Europe, Asia, and the United States. Several countries have achieved relatively high levels of wind power penetration, such as 21% of stationary electricity production in Denmark, 18% in Portugal, 16% in Spain, 14% in Ireland and 9% in Germany in 2010. As of 2011, 83 countries around the world are using wind power on a commercial basis.

As of 2012, the Alta Wind Energy Centre (California, 1,020 MW) is the world's largest wind farm. The London Array (630 MW) is the largest offshore wind farm in the world. The United Kingdom is the world's leading generator of offshore wind power, followed by Denmark.

There are many large wind farms under construction and these include Anholt Offshore Wind Farm (400 MW), BARD Offshore 1 (400 MW), Clyde Wind Farm (548 MW), Fβntβnele-Cogealac Wind Farm (600 MW), Greater Gabbard wind farm (500 MW), Lincs Wind Farm (270 MW), London Array (1000 MW), Lower Snake River Wind Project (343 MW), Macarthur Wind Farm (420 MW), Shepherds Flat Wind Farm (845 MW), and the Sheringham Shoal (317 MW).

Table: *Top 10 windpower countries of 2012*

Country	*capacity(MW)*	*% of total*
China	75,564	26.8
United States	60,007	21.2
Germany	31,332	11.1
Spain	22,796	8.1
India	18,421	6.5
United Kingdom	8,845	3.0
Italy	8,144	2.9

Contd...

Country	*capacity(MW)*	*% of total*
France	7,196	2.5
Canada	6,200	2.2
Portugal	4,525	1.6
(rest of world)	39,853	14.1
World total	282,482	100%

nameplate capacity is shown, figures by year-end.

Solar Thermal

The United States conducted much early research in photovoltaics and concentrated solar power. The U.S. is among the top countries in the world in electricity generated by the Sun and several of the world's largest utility-scale installations are located in the desert Southwest.

The oldest solar thermal power plant in the world is the 354 megawatt (MW) SEGS thermal power plant, in California. The Ivanpah Solar Electric Generating System is a solar thermal power project in the California Mojave Desert, 40 miles (64 km) southwest of Las Vegas, with a gross capacity of 377 MW. The 280 MW Solana Generating Station is a solar power plant near Gila Bend, Arizona, about 70 miles (110 km) southwest of Phoenix, completed in 2013. When commissioned it was the largest parabolic trough plant in the world and the first U.S. solar plant with molten salt thermal energy storage.

The solar thermal power industry is growing rapidly with 1.3 GW under construction in 2012 and more planned. Spain is the epicentre of solar thermal power development with 873 MW under construction, and a further 271 MW under development. In the United States, 5,600 MW of solar thermal power projects have been announced. Several power plants have been constructed in the Mojave Desert, Southwestern United States. The Ivanpah Solar Power Facility being the most recent. In developing countries, three World Bank projects for integrated solar thermal/combined-cycle gas-turbine power plants in Egypt, Mexico, and Morocco have been approved.

Photovoltaic Development

Photovoltaics (PV) uses solar cells assembled into solar panels to convert sunlight into electricity. It's a fast-growing technology doubling its worldwide installed capacity every couple of years. PV systems range from small, residential and commercial rooftop or building integrated installations, to large utility-scale solar plants. The predominant PV technology is crystalline silicon, while thin-film solar cell technology accounts for about 10 percent of global photovoltaic deployment. In recent years, PV technology has improved its electricity generating efficiency, reduced the installation cost per watt as well as its energy payback time (EPBT), and has reached grid parity in at least 19 different markets by 2014. Financial institutions are predicting a second solar "gold rush" in the near future.

At the end of 2013, worldwide PV capacity reached 139,000 megawatts. Photovoltaics grew fastest in China (+11.8 GW), followed by Japan (+6.9 GW) and the United States (+4.75 GW), while Germany remains the world's largest overall producer of photovoltaic power with a total capacity of 35.5 GW, contributing almost 6 percent to the overall electricity generation. Italy meets 7 percent of its electricity demands with photovoltaic power—the highest share worldwide.

For 2014, global photovoltaic capacity is estimated to increase by another 45 gigawatts (GW). By 2018, worldwide capacity is projected to reach as much as 430 gigawatts. This corresponds to a tripling within five years. Solar power is expected to become the world's largest source of electricity by 2050, with solar photovoltaics (PV) and solar thermal (CSP) contributing 16% and 11% respectively. This will require an increase of installed PV capacity from 139 GW to 4,600 GW, of which more than half will be deployed in China and India.

Photovoltaic Power Stations

Many solar photovoltaic power stations have been built, mainly in Europe. As of May 2012, the largest photovoltaic (PV) power plants in the world are the Agua Caliente Solar Project (USA, 247 MW), Charanka Solar Park (India, 214 MW), Golmud Solar Park (China, 200 MW), Perovo Solar Park (Ukraine, 100 MW), Sarnia Photovoltaic Power Plant (Canada, 97 MW), Brandenburg-Briest Solarpark (Germany, 91 MW), Solarpark Finow Tower (Germany, 84.7 MW), Montalto di Castro Photovoltaic Power Station (Italy, 84.2 MW), and the Eggebek Solar Park (Germany, 83.6 MW).

There are also many large plants under construction. The Desert Sunlight Solar Farm is a 550 MW solar power plant under construction in Riverside County, California, that will use thin-film solar photovoltaic modules made by First Solar. The Topaz Solar Farm is a 550 MW photovoltaic power plant, being built in San Luis Obispo County, California. The Blythe Solar Power Project is a 500 MW photovoltaic station under construction in Riverside County, California. The California Valley Solar Ranch (CVSR) is a 250 MW solar photovoltaic power plant, which is being built by SunPower in the Carrizo Plain, northeast of California Valley. The 230 MW Antelope Valley Solar Ranch is a First Solar photovoltaic project which is under construction in the Antelope Valley area of the Western Mojave Desert, and due to be completed in 2013.

Many of these plants are integrated with agriculture and some use tracking systems that follow the sun's daily path across the sky to generate more electricity than fixed-mounted systems. There are no fuel costs or emissions during operation of the power stations.

However, when it comes to renewable energy systems and PV, it is not just large systems that matter. Building-integrated photovoltaics or "onsite" PV systems use existing land and structures and generate power close to where it is consumed.

Carbon-neutral and Negative Fuels

Carbon-neutral fuels are synthetic fuels (including methane, gasoline, diesel fuel, jet fuel or ammonia) produced by hydrogenating waste carbon dioxide recycled from power plant flue-gas emissions, recovered from automotive exhaust gas, or derived from carbonic acid in seawater. Such fuels are considered carbon-neutral because they do not result in a net increase in atmospheric greenhouse gases. To the extent that synthetic fuels displace fossil fuels, or if they are produced from waste carbon or seawater carbonic acid, and their combustion is subject to carbon capture at the flue or exhaust pipe, they result in negative carbon dioxide emission and net carbon dioxide removal from the atmosphere, and thus constitute a form of greenhouse gas remediation.

Such renewable fuels alleviate the costs and dependency issues of imported fossil fuels without requiring either electrification of the vehicle fleet or conversion to hydrogen or other fuels, enabling continued compatible and affordable vehicles. Carbon-neutral fuels offer relatively low cost energy storage, alleviating the problems of wind and solar intermittency, and they enable distribution of wind, water, and solar power through existing natural gas pipelines. Nighttime wind power is considered the most economical form of electrical power with which to synthesize fuel, because the load curve for electricity peaks sharply during the warmest hours of the day, but wind tends to blow slightly more at night than during the day, so, the price of night-time wind power is often much less expensive than any alternative. Germany has built a 250 kilowatt synthetic methane plant which they are scaling up to 10 megawatts.

The George Olah carbon dioxide recycling plant in Grindavvk, Iceland has been producing 2 million liters of methanol transportation fuel per year from flue exhaust of the Svartsengi Power Station since 2011. It has the capacity to produce 5 million liters per year.

Biofuel Development

Biofuels provided 3% of the world's transport fuel in 2010. Mandates for blending biofuels exist in 31 countries at the national level and in 29 states/provinces. According to the International Energy Agency, biofuels have the potential to meet more than a quarter of world demand for transportation fuels by 2050.

Since the 1970s, Brazil has had an ethanol fuel program which has allowed the country to become the world's second largest producer of ethanol (after the United States) and the world's largest exporter. Brazil's ethanol fuel program uses modern equipment and cheap sugarcane as feedstock, and the residual cane-waste (bagasse) is used to produce heat and power. There are no longer light vehicles in Brazil running on pure gasoline. By the end of 2008 there were 35,000 filling stations throughout Brazil with at least one ethanol pump.

Nearly all the gasoline sold in the United States today is mixed with 10% ethanol, a mix known as E10, and motor vehicle manufacturers already produce vehicles designed to run on much higher ethanol blends. Ford, Daimler AG, and GM are among the automobile companies that sell "flexible-fuel" cars, trucks, and minivans that can use gasoline and ethanol blends ranging from pure gasoline up to 85% ethanol (E85). By mid-2006, there were approximately 6 million E85-compatible vehicles on U.S. roads. The challenge is to expand the market for biofuels beyond the farm states where they have been most popular to date. Flex-fuel vehicles are assisting in this transition because they allow drivers to choose different fuels based on price and availability. The *Energy Policy Act of 2005*, which calls for 7.5 billion US gallons (28,000,000 m^3) of biofuels to be used annually by 2012, will also help to expand the market.

Geothermal Development

Geothermal power is cost effective, reliable, sustainable, and environmentally friendly, but has historically been limited to areas near tectonic plate boundaries. Recent technological advances have expanded the range and size of viable resources, especially for applications such as home heating, opening a potential for widespread exploitation. Geothermal wells release greenhouse gases trapped deep within the earth, but these emissions are much lower per energy unit than those of fossil fuels. As a result, geothermal power has the potential to help mitigate global warming if widely deployed in place of fossil fuels.

The International Geothermal Association (IGA) has reported that 10,715 MW of geothermal power in 24 countries is online, which is expected to generate 67,246 GWh of electricity in 2010. This represents a 20% increase in geothermal power online capacity since 2005. IGA projects this will grow to 18,500 MW by 2015, due to the large number of projects presently under consideration, often in areas previously assumed to have little exploitable resource.

In 2010, the United States led the world in geothermal electricity production with 3,086 MW of installed capacity from 77 power plants; the largest group of geothermal power plants in the world is located at The Geysers, a geothermal field in California. The Philippines follows the US as the second highest producer of geothermal power in the world, with 1,904 MW of capacity online; geothermal power makes up approximately 18% of the country's electricity generation.

Developing Countries

Renewable energy can be particularly suitable for developing countries. In rural and remote areas, transmission and distribution of energy generated from fossil fuels can be difficult and expensive. Producing renewable energy locally can offer a viable alternative.

Technology advances are opening up a huge new market for solar power: the approximately 1.3 billion people around the world who don't have access to grid electricity. Even though they are typically very poor, these people have to pay far more for lighting than people in rich countries because they use inefficient kerosene lamps. Solar power costs half as much as lighting with kerosene. An estimated 3 million households get power from small solar PV systems. Kenya is the world leader in the number of solar power systems installed per capita. More than 30,000 very small solar panels, each producing 12 to 30 watts, are sold in Kenya annually. Some Small Island Developing States (SIDS) are also turning to solar power to reduce their costs and increase their sustainability.

Micro-hydro configured into mini-grids also provide power. Over 44 million households use biogas made in household-scale digesters for lighting and/or cooking, and more than 166 million households rely on a new generation of more-efficient biomass cookstoves. Clean liquid fuel sourced from renewable feedstocks are used for cooking and lighting in energy-poor areas of the developing world. Alcohol fuels (ethanol and methanol) can be produced sustainably from non-food sugary, starchy, and cellulostic feedstocks. Project Gaia, Inc. and CleanStar Mozambique are implementing clean cooking programs with liquid ethanol stoves in Ethiopia, Kenya, Nigeria and Mozambique.

Renewable energy projects in many developing countries have demonstrated that renewable energy can directly contribute to poverty reduction by providing the energy needed for creating businesses and employment. Renewable energy technologies can also make indirect contributions to alleviating poverty by providing energy for cooking, space heating, and lighting. Renewable energy can also contribute to education, by providing electricity to schools.

Industry and Policy Trends

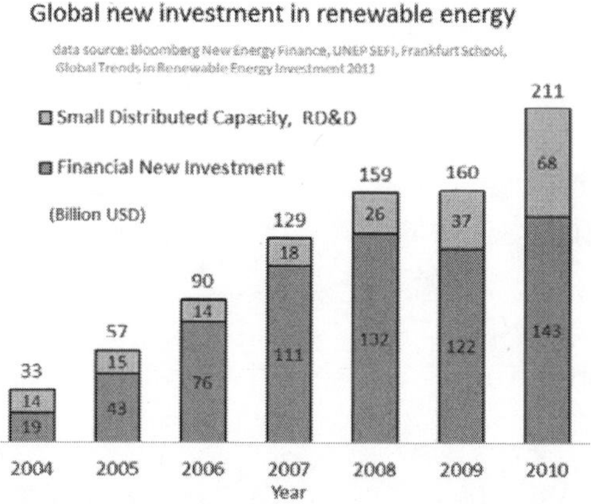

Figure: *Global New Investments in Renewable Energy*

U.S. President Barack Obama's American Recovery and Reinvestment Act of 2009 includes more than $70 billion in direct spending and tax credits for clean energy and associated transportation programs. Clean Edge suggests that the commercialization of clean energy will help countries around the world pull out of the current economic malaise. Leading renewable energy companies include First Solar, Gamesa, GE Energy, Q-Cells, Sharp Solar, Siemens, SunOpta, Suntech Power, and Vestas.

The military has also focused on the use of renewable fuels for military vehicles. Unlike fossil fuels, renewable fuels can be produced in any country, creating a strategic advantage. The US military has already committed itself to have 50% of its energy consumption come from alternative sources. The International Renewable Energy Agency (IRENA) is an intergovernmental organization for promoting the adoption of renewable energy worldwide. It aims to provide concrete policy advice and facilitate capacity building and technology transfer.

IRENA was formed on January 26, 2009, by 75 countries signing the charter of IRENA. As of March 2010, IRENA has 143 member states who all are considered as founding members, of which 14 have also ratified the statute.

As of 2011, 119 countries have some form of national renewable energy policy target or renewable support policy. National targets now exist in at least 98 countries. There is also a wide range of policies at state/provincial and local levels.

United Nations' Secretary-General Ban Ki-moon has said that renewable energy has the ability to lift the poorest nations to new levels of prosperity. In October 2011, he "announced the creation of a high-level group to drum up support for energy access, energy efficiency and greater use of renewable energy. The group is to be co-chaired by Kandeh Yumkella, the chair of UN Energy and director general of the UN Industrial Development Organisation, and Charles Holliday, chairman of Bank of America".

100% Renewable Energy

The incentive to use 100% renewable energy, for electricity, transport, or even total primary energy supply globally, has been motivated by global warming and other ecological as well as economic concerns. The Intergovernmental Panel on Climate Change has said that there are few fundamental technological limits to integrating a portfolio of renewable energy technologies to meet most of total global energy demand. Renewable energy use has grown much faster than even advocates anticipated.

At the national level, at least 30 nations around the world already have renewable energy contributing more than 20% of energy supply. Also, Professors S. Pacala and Robert H. Socolow have developed a series of "stabilization wedges" that can allow us to maintain our quality of life while avoiding catastrophic climate change, and "renewable energy sources," in aggregate, constitute the largest number of their "wedges."

Mark Z. Jacobson, professor of civil and environmental engineering at Stanford University and director of its Atmosphere and Energy Program says producing all new energy with wind power, solar power, and hydropower by 2030 is feasible and existing energy supply arrangements could be replaced by 2050. Barriers to implementing the renewable energy plan are seen to be "primarily social and political, not technological or economic". Jacobson says that energy costs with a wind, solar, water system should be similar to today's energy costs.

Similarly, in the United States, the independent National Research Council has noted that "sufficient domestic renewable resources exist to allow renewable electricity to play a significant role in future electricity generation and thus help confront issues related to climate change, energy security, and the escalation of energy costs ... Renewable energy is an attractive option because renewable resources available in the United States, taken collectively, can supply significantly greater amounts of electricity than the total current or projected domestic demand." .

The most significant barriers to the widespread implementation of large-scale renewable energy and low carbon energy strategies are primarily political and not technological. According to the 2013 *Post Carbon Pathways* report, which reviewed many international studies, the key roadblocks are: climate change denial, the fossil fuels lobby, political inaction, unsustainable energy consumption, outdated energy infrastructure, and financial constraints.

Coal

Coal (from the Old English term *col*, which has meant "mineral of fossilized carbon" since the 13th century) is a combustible black or brownish-black sedimentary rock usually occurring in rock strata in layers or veins called coal beds or coal seams. The harder forms, such as anthracite coal, can be regarded as metamorphic rock because of later exposure to elevated temperature and pressure. Coal is composed primarily of carbon along with variable quantities of other elements, chiefly hydrogen, sulfur, oxygen, and nitrogen.

Throughout history, coal has been used as an energy resource, primarily burned for the production of electricity and/or heat, and is also used for industrial purposes, such as refining metals. A fossil fuel, coal forms when dead plant matter is converted into peat, which in turn is converted into lignite, then sub-bituminous coal, after that bituminous coal, and lastly anthracite. This involves biological and geological processes that take place over a long period. The Energy Information Administration estimates coal reserves at 948×10^9 short tons (860 Gt). One estimate for resources is 18 000 Gt.

Coal is the largest source of energy for the generation of electricity worldwide, as well as one of the largest worldwide anthropogenic sources of carbon dioxide releases. In 1999, world gross carbon dioxide emissions from coal usage were 8,666 million tonnes of carbon dioxide. In 2011, world gross emissions from coal usage were 14,416 million tonnes. Coal-fired electric power generation emits around 2,000 pounds of carbon dioxide for every megawatt-hour generated, which is almost double the approximately 1100 pounds of carbon dioxide released by a natural gas-fired electric plant per megawatt-hour generated. Because of this higher carbon efficiency of natural gas generation, as the market in the United States has changed to reduce coal and increase natural gas generation, carbon dioxide emissions have fallen. Those measured in the first quarter of 2012 were the lowest of any recorded for the first quarter of any year since 1992. In 2013, the head of the UN climate agency advised that most of the world's coal reserves should be left in the ground to avoid catastrophic global warming.

Coal is extracted from the ground by coal mining, either underground by shaft mining, or at ground level by open pit mining extraction. Since 1983 the world top coal producer has been China. In 2011 China produced 3,520 million tonnes of coal – 49.5% of 7,695 million tonnes world coal production. In 2011 other large producers were United States (993 million tonnes), India (589), European Union (576) and Australia (416). In 2010 the largest exporters were Australia with 328 million tonnes (27.1% of world coal export) and Indonesia with 316 million tonnes (26.1%), while the largest importers were Japan with 207 million tonnes (17.5% of world coal import), China with 195 million tonnes (16.6%) and South Korea with 126 million tonnes (10.7%).

Types

As geological processes apply pressure to dead biotic material over time, under suitable conditions it is transformed successively into:

- Peat, considered to be a precursor of coal, has industrial importance as a fuel in some regions, for example, Ireland and Finland. In its dehydrated form, peat is a highly effective absorbent for fuel and oil spills on land and water. It is also used as a conditioner for soil to make it more able to retain and slowly release water.
- Lignite, or brown coal, is the lowest rank of coal and used almost exclusively as fuel for electric power generation. Jet, a compact form of lignite, is sometimes polished and has been used as an ornamental stone since the Upper Palaeolithic.
- Sub-bituminous coal, whose properties range from those of lignite to those of bituminous coal, is used primarily as fuel for steam-electric power generation and is an important source of light aromatic hydrocarbons for the chemical synthesis industry.
- Bituminous coal is a dense sedimentary rock, usually black, but sometimes dark brown, often with well-defined bands of

bright and dull material; it is used primarily as fuel in steam-electric power generation, with substantial quantities used for heat and power applications in manufacturing and to make coke.

- "Steam coal" is a grade between bituminous coal and anthracite, once widely used as a fuel for steam locomotives. In this specialized use, it is sometimes known as "sea-coal" in the US. Small steam coal (dry small steam nuts or DSSN) was used as a fuel for domestic water heating.
- Anthracite, the highest rank of coal, is a harder, glossy black coal used primarily for residential and commercial space heating. It may be divided further into metamorphically altered bituminous coal and "petrified oil", as from the deposits in Pennsylvania.
- Graphite, technically the highest rank, is difficult to ignite and is not commonly used as fuel — it is mostly used in pencils and, when powdered, as a lubricant.

The classification of coal is generally based on the content of volatiles. However, the exact classification varies between countries. According to the German classification, coal is classified as follows:

German Classification	***English Designation***	***Volatiles %***	***C Carbon %***	***H Hydrogen %***	***O Oxygen %***	***S Sulfur %***	***Heat content kJ/kg***
Braunkohle	Lignite (brown coal)	45–65	60–75	6.0–5.8	34-17	0.5-3	<28,470
Flammkohle	Flame coal	40-45	75-82	6.0-5.8	>9.8	~1	<32,870
Gasflammkohle	Gas flame coal	35-40	82-85	5.8-5.6	9.8-7.3	~1	<33,910
Gaskohle	Gas coal	28-35	85-87.5	5.6-5.0	7.3-4.5	~1	<34,960
Fettkohle	Fat coal	19-28	87.5-89.5	5.0-4.5	4.5-3.2	~1	<35,380
Esskohle	Forge coal	14-19	89.5-90.5	4.5-4.0	3.2-2.8	~1	<35,380
Magerkohle	Nonbaking coal	10-14	90.5-91.5	4.0-3.75	2.8-3.5	~1	35,380
Anthrazit	Anthracite	7-12	>91.5	<3.75	<2.5	~1	<35,300
Note, the percentages are percent by mass of the indicated elements							

The middle six grades in the table represent a progressive transition from the English-language sub-bituminous to bituminous coal, while the last class is an approximate equivalent to anthracite, but more inclusive (US anthracite has < 6% volatiles).

Cannel coal (sometimes called "candle coal") is a variety of fine-grained, high-rank coal with significant hydrogen content. It consists primarily of "exinite" macerals, now termed "liptinite".

Coal as Fuel

Coal is primarily used as a solid fuel to produce electricity and heat through combustion. World coal consumption was about 7.25 billion tonnes in 2010 (7.99 billion short tons) and is expected to increase 48% to 9.05 billion tonnes (9.98 billion short tons) by 2030. China produced 3.47 billion tonnes (3.83 billion short tons) in 2011. India produced about 578 million tonnes (637.1 million short tons) in 2011. 68.7% of China's electricity comes from coal. The USA consumed about 13% of the world total in 2010, i.e. 951 million tonnes (1.05 billion short tons), using 93% of it for generation of electricity. 46% of total power generated in the USA was done using coal.

When coal is used for electricity generation, it is usually pulverized and then combusted (burned) in a furnace with a boiler. The furnace heat converts boiler water to steam, which is then used to spin turbines which turn generators and create electricity. The thermodynamic efficiency of this process has been improved over time; some older coal-fired power stations have thermal efficiencies in the vicinity of 25% whereas the newest supercritical and "ultra-supercritical" steam cycle turbines, operating at temperatures over 600 °C and pressures over 27 MPa (over 3900 psi), can practically achieve thermal efficiencies in excess of 45% (LHV basis) using anthracite fuel, or around 43% (LHV basis) even when using lower-grade lignite fuel. Further thermal efficiency improvements are also achievable by improved pre-drying (especially relevant with high-moisture fuel such as lignite or biomass) and cooling technologies.

An alternative approach of using coal for electricity generation with improved efficiency is the integrated gasification combined cycle (IGCC) power plant. Instead of pulverizing the coal and burning it directly as fuel in the steam-generating boiler, the coal can be first gasified to create syngas, which is burned in a gas turbine to produce electricity (just like natural gas is burned in a turbine). Hot exhaust gases from the turbine are used to raise steam in a heat recovery steam generator which powers a supplemental steam turbine. Thermal efficiencies of current IGCC power plants range from 39-42% (HHV basis) or ~42-45% (LHV basis) for bituminous coal and assuming utilization of mainstream gasification technologies (Shell, GE Gasifier, CB&I). IGCC power plants outperform conventional pulverized coal-fueled plants in terms of pollutant emissions, and allow for relatively easy carbon capture.

At least 40% of the world's electricity comes from coal, and in 2012, about one-third of the United States' electricity came from coal, down from approximately 49% in 2008. As of 2012 in the United States, use of coal to generate electricity was declining, as plentiful supplies of natural gas obtained by hydraulic fracturing of tight shale formations became available at low prices.

In Denmark, a net electric efficiency of > 47% has been obtained at the coal-fired Nordjyllandsværket CHP Plant and an overall plant efficiency of up to 91% with cogeneration of electricity and district heating. The multifuel-fired Avedøreværket CHP Plant just outside Copenhagen can achieve a net electric efficiency as high as 49%. The overall plant efficiency with cogeneration of electricity and district heating can reach as much as 94%.

An alternative form of coal combustion is as coal-water slurry fuel (CWS), which was developed in the Soviet Union. CWS significantly reduces emissions, improving the heating value of coal. Other ways to use coal are combined heat and power cogeneration and an MHD topping cycle.

The total known deposits recoverable by current technologies, including highly polluting, low-energy content types of coal (i.e., lignite, bituminous), is sufficient for many years. However, consumption is increasing and maximal production could be reached within decades. On the other hand much may have to be left in the ground to avoid climate change.

Coking Coal and Use of Coke

Coke is a solid carbonaceous residue derived from low-ash, low-sulfur bituminous coal from which the volatile constituents are driven off by baking in an oven without oxygen at temperatures as high as 1,000 °C (1,832 °F), so the fixed carbon and residual ash are fused together. Metallurgical coke is used as a fuel and as a reducing agent in smelting iron ore in a blast furnace. The result is pig iron, and is too rich in dissolved carbon, so it must be treated further to make steel. The coking coal should be low in sulfur and phosphorus, so they do not migrate to the metal.

The coke must be strong enough to resist the weight of overburden in the blast furnace, which is why coking coal is so important in making steel using the conventional route. However, the alternative route is direct reduced iron, where any carbonaceous fuel can be used to make sponge or pelletised iron. Coke from coal is grey, hard, and porous and has a heating value of 24.8 million Btu/ton (29.6 MJ/kg). Some cokemaking processes produce valuable byproducts, including coal tar, ammonia, light oils, and coal gas.

Petroleum coke is the solid residue obtained in oil refining, which resembles coke, but contains too many impurities to be useful in metallurgical applications.

Gasification

Coal gasification can be used to produce syngas, a mixture of carbon monoxide (CO) and hydrogen (H_2) gas. Often syngas is used to fire gas turbines to produce electricity, but the versatility of syngas also allows it to be converted into transportation fuels, such as gasoline and diesel, through the Fischer-Tropsch process; alternatively, syngas can be converted into methanol, which can be blended into fuel directly or converted to gasoline via the methanol to gasoline process. Gasification combined with Fischer-Tropsch technology is currently used by the Sasol chemical company of South Africa to make motor vehicle fuels from coal and natural gas. Alternatively, the hydrogen obtained from gasification can be used for various purposes, such as powering a hydrogen economy, making ammonia, or upgrading fossil fuels.

During gasification, the coal is mixed with oxygen and steam while also being heated and pressurized. During the reaction, oxygen and water molecules oxidize the coal into carbon monoxide (CO), while also releasing hydrogen gas (H_2). This process has been conducted in both underground coal mines and in the production of town gas.

$$C\ (\textit{as Coal}) + O_2 + H_2O \rightarrow H_2 + CO$$

If the refiner wants to produce gasoline, the syngas is collected at this state and routed into a Fischer-Tropsch reaction. If hydrogen is the desired end-product, however, the syngas is fed into the water gas shift reaction, where more hydrogen is liberated.

$$CO + H_2O \rightarrow CO_2 + H_2$$

In the past, coal was converted to make coal gas (town gas), which was piped to customers to burn for illumination, heating, and cooking.

Liquefaction

Coal can also be converted into synthetic fuels equivalent to gasoline or diesel by several different direct processes (which do not intrinsically require gasification or indirect conversion). In the direct liquefaction processes, the coal is either hydrogenated or carbonized. Hydrogenation processes are the Bergius process, the SRC-I and SRC-

II (Solvent Refined Coal) processes, the NUS Corporation hydrogenation process and several other single-stage and two-stage processes. In the process of low-temperature carbonization, coal is coked at temperatures between 360 and 750 °C (680 and 1,380 °F). These temperatures optimize the production of coal tars richer in lighter hydrocarbons than normal coal tar. The coal tar is then further processed into fuels. An overview of coal liquefaction and its future potential is available.

Coal liquefaction methods involve carbon dioxide (CO_2) emissions in the conversion process. If coal liquefaction is done without employing either carbon capture and storage (CCS) technologies or biomass blending, the result is lifecycle greenhouse gas footprints that are generally greater than those released in the extraction and refinement of liquid fuel production from crude oil.

If CCS technologies are employed, reductions of 5–12% can be achieved in Coal to Liquid (CTL) plants and up to a 75% reduction is achievable when co-gasifying coal with commercially demonstrated levels of biomass (30% biomass by weight) in coal/biomass-to-liquids plants. For future synthetic fuel projects, carbon dioxide sequestration is proposed to avoid releasing CO_2 into the atmosphere. Sequestration adds to the cost of production.

Refined Coal

Refined coal is the product of a coal-upgrading technology that removes moisture and certain pollutants from lower-rank coals such as sub-bituminous and lignite (brown) coals. It is one form of several precombustion treatments and processes for coal that alter coal's characteristics before it is burned. The goals of precombustion coal technologies are to increase efficiency and reduce emissions when the coal is burned. Depending on the situation, precombustion technology can be used in place of or as a supplement to postcombustion technologies to control emissions from coal-fuelled boilers.

Industrial Processes

Finely ground bituminous coal, known in this application as sea coal, is a constituent of foundry sand. While the molten metal is in the mould, the coal burns slowly, releasing reducing gases at pressure, and so preventing the metal from penetrating the pores of the sand. It is also contained in 'mould wash', a paste or liquid with the same function applied to the mould before casting. Sea coal can be mixed with the clay lining (the "bod") used for the bottom of a cupola furnace. When heated, the coal decomposes and the bod becomes slightly friable, easing the process of breaking open holes for tapping the molten metal.

Production of Chemicals

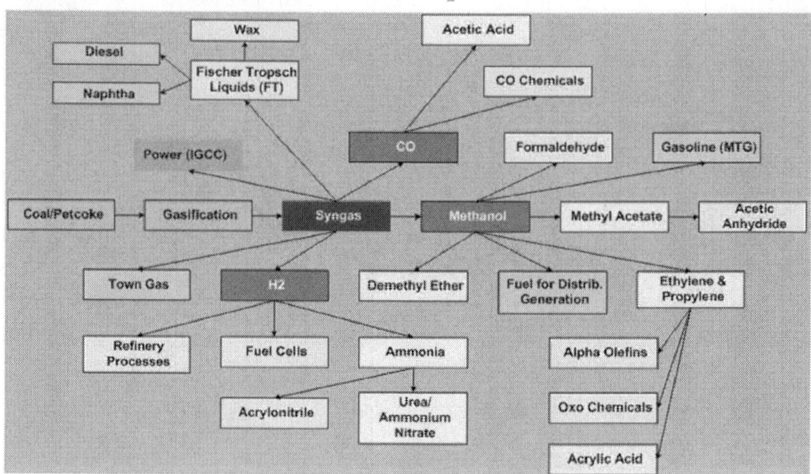

Figure: *Production of Chemicals from Coal*

Coal is an important feedstock in production of a wide range of chemical fertilizers and other chemical products. The main route to these products is coal gasification to produce syngas. Primary chemicals that are produced directly from the syngas include methanol, hydrogen and carbon monoxide, which are the chemical building blocks from which a whole spectrum of derivative chemicals are manufactured, including olefins, acetic acid, formaldehyde, ammonia, urea and others. The versatility of syngas as a precursor to primary chemicals and high-value derivative products provides the option of using relatively inexpensive coal to produce a wide range of valuable commodities.

Historically, production of chemicals from coal has been used since the 1950s and has become established in the market. According to the 2010 Worldwide Gasification Database, a survey of current and planned gasifiers, from 2004 to 2007 chemical production increased its gasification product share from 37% to 45%. From 2008 to 2010, 22% of new gasifier additions were to be for chemical production.

Because the slate of chemical products that can be made via coal gasification can in general also use feedstocks derived from natural gas and petroleum, the chemical industry tends to use whatever feedstocks are most cost-effective. Therefore, interest in using coal tends to increase for higher oil and natural gas prices and during periods of high global economic growth that may strain oil and gas production. Also, production of chemicals from coal is of much higher interest in countries like South Africa, China, India and the United States where there are abundant coal resources. The abundance of coal combined with lack of natural gas resources in China is strong inducement for the coal to chemicals industry pursued there. In the United States, the best example of the industry is Eastman Chemical Company which has been successfully operating a coal-to-chemicals plant at its Kingsport, Tennessee, site since 1983. Similarly, Sasol has built and operated coal-to-chemicals facilities in South Africa.

Coal to chemical processes do require substantial quantities of water. As of 2013 much of the coal to chemical production was in the People's Republic of China where environmental regulation and water management was weak.

Coal as a Traded Commodity

In North America, Central Appalachian coal futures contracts are currently traded on the New York Mercantile Exchange (trading symbol *QL*). The trading unit is 1,550 short tons (1,410 t) per contract, and is quoted in U.S. dollars and cents per ton. Since coal is the principal fuel for generating electricity in the United States, coal futures contracts provide coal producers and the electric power industry an important tool for hedging and risk management.

In addition to the NYMEX contract, the Intercontinental Exchange (ICE) has European (Rotterdam) and South African (Richards Bay) coal futures available for trading. The trading unit for these contracts is 5,000 tonnes (5,500 short tons), and are also quoted in U.S. dollars and cents per ton.

The price of coal increased from around $30.00 per short ton in 2000 to around $150.00 per short ton as of September 2008. As of October 2008, the price per short ton had declined to $111.50. Prices further declined to $71.25 as of October 2010.

Economic Aspects

Coal (by liquefaction technology) is one of the backstop resources that could limit escalation of oil prices and mitigate the effects of transportation energy shortage that will occur under peak oil. This is contingent on liquefaction production capacity becoming large enough to satiate the very large and growing demand for petroleum. Estimates of the cost of producing liquid fuels from coal suggest that domestic U.S. production of fuel from coal becomes cost-competitive with oil priced at around $35 per barrel, with the $35 being the break-even cost. With oil prices as low as around $40 per barrel in the U.S. as of December 2008, liquid coal lost some of its economic allure in the U.S., but will probably be re-vitalized, similar to oil sand projects, with an oil price around $70 per barrel.

In China, due to an increasing need for liquid energy in the transportation sector, coal liquefaction projects were given high priority even during periods of oil prices below $40 per barrel. This is probably because China prefers not to be dependent on foreign oil, instead utilizing its enormous domestic coal reserves. As oil prices were increasing during the first half of 2009, the coal liquefaction projects in China were again boosted, and these projects are profitable with an oil barrel price of $40.

China is the largest producer of coal in the world. It is the world's largest energy consumer, and relies on coal to supply 69% of its energy needs. An estimated 5 million people worked in China's coal-mining industry in 2007.

Coal pollution costs the EU €43 billion each year. Measures to cut air pollution may have beneficial long-term economic impacts for individuals.

Energy Density and Carbon Impact

The energy density of coal, i.e. its heating value, is roughly 24 megajoules per kilogram (approximately 6.7 kilowatt-hours per kg). For a coal power plant with a 40% efficiency, it takes an estimated 325 kg (717 lb) of coal to power a 100 W lightbulb for one year.

As of 2006, the average efficiency of electricity-generating power stations was 31%; in 2002, coal represented about 23% of total global energy supply, an equivalent of 3.4 billion tonnes of coal, of which 2.8 billion tonnes were used for electricity generation.

The US Energy Information Agency's 1999 report on CO_2 emissions for energy generation quotes an emission factor of 0.963 kg CO_2/kWh for coal power, compared to 0.881 kg CO_2/kWh (oil), or 0.569 kg CO_2/kWh (natural gas).

Production Trends

In 2006, China was the top producer of coal with 38% share followed by the United States and India, according to the British Geological Survey. As of 2012 coal production in the United States was falling at the rate of 7% annually with many power plants using coal shut down or converted to natural gas; however, some of the reduced domestic demand was taken up by increased exports with five coal export terminals being proposed in the Pacific Northwest to export coal from the Powder River Basin to China and other Asian markets; however, as of 2013, environmental opposition was increasing. High-sulfur coal mined in Illinois which was unsaleable in the United States found a ready market in Asia as exports reached 13 million tons in 2012.

World Coal Reserves

The 948 billion short tons of recoverable coal reserves estimated by the Energy Information Administration are equal to about 4,196 BBOE (billion barrels of oil equivalent). The amount of coal burned during 2007 was estimated at 7.075 billion short tons, or 133.179 quadrillion BTU's. This is an average of 18.8 million BTU per short ton. In terms of heat content, this is about 57,000,000 barrels (9,100,000 m) of oil equivalent per day. By comparison in 2007, natural gas provided 51,000,000 barrels (8,100,000 m^3) of oil equivalent per day, while oil provided 85,800,000 barrels (13,640,000 m^3) per day.

British Petroleum, in its 2007 report, estimated at 2006 end that there were 147 years reserves-to-production ratio based on *proven* coal reserves worldwide. This figure only includes reserves classified as "proven"; exploration drilling programs by mining companies, particularly in under-explored areas, are continually providing new reserves. In many cases, companies are aware of coal deposits that have not been sufficiently drilled to qualify as "proven". However, some nations haven't updated their information and assume reserves remain at the same levels even with withdrawals.

Petroleum

Petroleum (L. *petroleum*, from early 15c. "petroleum, rock oil" (mid-14c. in Anglo-French), from Medieval Latin petroleum, from Latin petra *rock* + Latin: *oleum oil* is a naturally occurring, yellow-to-black liquid found in geologic formations beneath the Earth's surface, which is commonly refined into various types of fuels. It consists of hydrocarbons of various molecular weights and other organic compounds. The name *petroleum* covers both naturally occurring unprocessed crude oil and petroleum products that are made up of refined crude oil. A fossil fuel, petroleum is formed when large quantities of dead organisms, usually zooplankton and algae, are buried underneath sedimentary rock and subjected to intense heat and pressure.

Petroleum is recovered mostly through oil drilling (natural petroleum springs are rare). This comes after the studies of structural geology (at the reservoir scale), sedimentary basin analysis, reservoir characterization (mainly in terms of the porosity and permeability of geologic reservoir structures). It is refined and separated, most easily by distillation, into a large number of consumer products, from gasoline (petrol) and kerosene to asphalt and chemical reagents used to make plastics and pharmaceuticals. Petroleum is used in manufacturing a wide variety of materials, and it is estimated that the world consumes about 90 million barrels each day. The use of fossil fuels, such as petroleum, has a negative impact on Earth's biosphere, releasing pollutants and greenhouse gases into the air and damaging ecosystems through events such as oil spills. Concern over the depletion of the earth's finite reserves of oil, and the effect this would have on a society dependent on it, is a concept known as peak oil.

Modern History

In 1847, the process to distill kerosene from petroleum was invented by James Young. He noticed a natural petroleum seepage in the Riddings colliery at Alfreton, Derbyshire from which he distilled a light thin oil suitable for use as lamp oil, at the same time obtaining a thicker oil suitable for lubricating machinery. In 1848 Young set up a small business refining the crude oil.

Young eventually succeeded, by distilling cannel coal at a low heat, in creating a fluid resembling petroleum, which when treated in the same way as the seep oil gave similar products. Young found that by slow distillation he could obtain a number of useful liquids from it, one of which he named "paraffine oil" because at low temperatures it congealed into a substance resembling paraffin wax.

The production of these oils and solid paraffin wax from coal formed the subject of his patent dated 17 October 1850. In 1850 Young & Meldrum and Edward William Binney entered into partnership under the title of E.W. Binney & Co. at Bathgate in West Lothian and E. Meldrum & Co. at Glasgow; their works at Bathgate were completed in 1851 and became the first truly commercial oil-works in the world with the first modern oil refinery, using oil extracted from locally-mined torbanite, shale, and bituminous coal to manufacture naphtha and lubricating oils; paraffin for fuel use and solid paraffin were not sold until 1856.

Another early refinery was built by Ignacy £ukasiewicz, providing a cheaper alternative to whale oil. The demand for petroleum as a fuel for lighting in North America and around the world quickly grew. Edwin Drake's 1859 well near Titusville, Pennsylvania, is popularly considered the first modern well. Drake's well is probably singled out because it was drilled, not dug; because it used a steam engine; because there was a company associated with it; and because it touched off a major boom. However, there was considerable activity before Drake in various parts of the world in the mid-19th century. A group directed by Major Alexeyev of the Bakinskii Corps of Mining Engineers hand-drilled a well in the Baku region in 1848. There were engine-drilled wells in

West Virginia in the same year as Drake's well. An early commercial well was hand dug in Poland in 1853, and another in nearby Romania in 1857. At around the same time the world's first, small, oil refinery was opened at Jas³o in Poland, with a larger one opened at Ploie'ti in Romania shortly after. Romania is the first country in the world to have had its annual crude oil output officially recorded in international statistics: 275 tonnes for 1857.

The first commercial oil well in Canada became operational in 1858 at Oil Springs, Ontario (then Canada West). Businessman James Miller Williams dug several wells between 1855 and 1858 before discovering a rich reserve of oil four metres below ground. Williams extracted 1.5 million litres of crude oil by 1860, refining much of it into kerosene lamp oil. William's well became commercially viable a year before Drake's Pennsylvania operation and could be argued to be the first commercial oil well in North America. The discovery at Oil Springs touched off an oil boom which brought hundreds of speculators and workers to the area. Advances in drilling continued into 1862 when local driller Shaw reached a depth of 62 metres using the spring-pole drilling method. On January 16, 1862, after an explosion of natural gas Canada's first oil gusher came into production, shooting into the air at a recorded rate of 3,000 barrels per day. By the end of the 19th century the Russian Empire, particularly the Branobel company in Azerbaijan, had taken the lead in production.

Access to oil was and still is a major factor in several military conflicts of the twentieth century, including World War II, during which oil facilities were a major strategic asset and were extensively bombed. The German invasion of the Soviet Union included the goal to capture the Baku oilfields, as it would provide much needed oil-supplies for the German military which was suffering from blockades. Oil exploration in North America during the early 20th century later led to the US becoming the leading producer by mid-century. As petroleum production in the US peaked during the 1960s, however, the United States was surpassed by Saudi Arabia and the Soviet Union.

Today, about 90 percent of vehicular fuel needs are met by oil. Petroleum also makes up 40 percent of total energy consumption in the United States, but is responsible for only 1 percent of electricity generation. Petroleum's worth as a portable, dense energy source powering the vast majority of vehicles and as the base of many industrial chemicals makes it one of the world's most important commodities. Viability of the oil commodity is controlled by several key parameters, number of vehicles in the world competing for fuel, quantity of oil exported to the world market (Export Land Model), Net Energy Gain (economically useful energy provided minus energy consumed), political stability of oil exporting nations and ability to defend oil supply lines.

The top three oil producing countries are Russia, Saudi Arabia and the United States. About 80 percent of the world's readily accessible reserves are located in the Middle East, with 62.5 percent coming from the Arab 5: Saudi Arabia, UAE, Iraq, Qatar and Kuwait. A large portion of the world's total oil exists as unconventional sources, such as bitumen in Canada and extra heavy oil in Venezuela. While significant volumes of oil are extracted from oil sands, particularly in Canada, logistical and technical hurdles remain, as oil extraction requires large amounts of heat and water, making its net energy content quite low relative to conventional crude oil. Thus, Canada's oil sands are not expected to provide more than a few million barrels per day in the foreseeable future.

Composition

In its strictest sense, petroleum includes only crude oil, but in common usage it includes all liquid, gaseous, and solid hydrocarbons. Under surface pressure and temperature conditions, lighter hydrocarbons methane, ethane, propane and butane occur as gases, while pentane and heavier ones are in the form of liquids or solids. However, in an underground oil reservoir the proportions of gas, liquid, and solid depend on subsurface conditions and on the phase diagram of the petroleum mixture.

An oil well produces predominantly crude oil, with some natural gas dissolved in it. Because the pressure is lower at the surface than underground, some of the gas will come out of solution and be recovered (or burned) as *associated gas* or *solution gas.* A gas well produces predominantly natural gas. However, because the underground temperature and pressure are higher than at the surface, the gas may contain heavier hydrocarbons such as pentane, hexane, and heptane in the gaseous state. At surface conditions these will condense out of the gas to form natural gas condensate, often shortened to *condensate.* Condensate resembles petrol in appearance and is similar in composition to some volatile light crude oils.

The proportion of light hydrocarbons in the petroleum mixture varies greatly among different oil fields, ranging from as much as 97 percent by weight in the lighter oils to as little as 50 percent in the heavier oils and bitumens.

The hydrocarbons in crude oil are mostly alkanes, cycloalkanes and various aromatic hydrocarbons while the other organic compounds contain nitrogen, oxygen and sulfur, and trace amounts of metals such as iron, nickel, copper and vanadium. The exact molecular composition varies widely from formation to formation but the proportion of chemical elements vary over fairly narrow limits as follows:

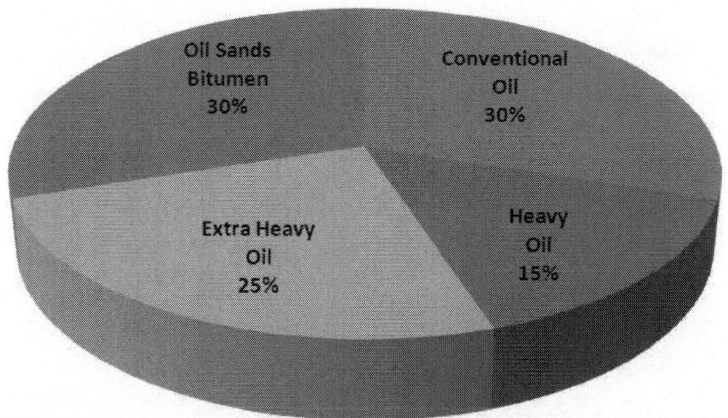

Composition by weight

Element	*Percent range*
Carbon	83 to 85%
Hydrogen	10 to 14%
Nitrogen	0.1 to 2%
Oxygen	0.05 to 1.5%
Sulfur	0.05 to 6.0%
Metals	< 0.1%

Four different types of hydrocarbon molecules appear in crude oil. The relative percentage of each varies from oil to oil, determining the properties of each oil.

Composition by weight

Hydrocarbon	*Average*	*Range*
Alkanes (paraffins)	30%	15 to 60%
Naphthenes	49%	30 to 60%
Aromatics	15%	3 to 30%
Asphaltics	6%	remainder

Crude oil varies greatly in appearance depending on its composition. It is usually black or dark brown (although it may be yellowish, reddish, or even greenish). In the reservoir it is usually found in association with natural gas, which being lighter forms a gas cap over the petroleum, and saline water which, being heavier than most forms of crude oil, generally sinks beneath it. Crude oil may also be found in semi-solid form mixed with sand and water, as in the Athabasca oil sands in Canada, where it is usually referred to as crude bitumen. In Canada, bitumen is considered a sticky, black, tar-like form of crude oil which is so thick and heavy that it must be heated or diluted before it will flow. Venezuela also has large amounts of oil in the Orinoco oil sands, although the hydrocarbons trapped in them are more fluid than in Canada and are usually called extra heavy oil. These oil sands resources are called unconventional oil to distinguish them from oil which can be extracted using traditional oil well methods. Between them, Canada and Venezuela contain an estimated 3.6 trillion barrels (570×10^9 m^3) of bitumen and extra-heavy oil, about twice the volume of the world's reserves of conventional oil.

Petroleum is used mostly, by volume, for producing fuel oil and petrol, both important *"primary energy"* sources. 84 percent by volume of the hydrocarbons present in petroleum is converted into energy-rich fuels (petroleum-based fuels), including petrol, diesel, jet, heating, and other fuel oils, and liquefied petroleum gas. The lighter grades of crude oil produce the best yields of these products, but as the world's reserves of light and medium oil are depleted, oil refineries are increasingly having to process heavy oil and bitumen, and use more complex and expensive methods to produce the products required. Because heavier crude oils have too much carbon and not enough hydrogen, these processes generally involve removing carbon from or adding hydrogen to the molecules, and using fluid catalytic cracking to convert the longer, more complex molecules in the oil to the shorter, simpler ones in the fuels.

Due to its high energy density, easy transportability and relative abundance, oil has become the world's most important source of energy since the mid-1950s. Petroleum is also the raw material for many chemical products, including pharmaceuticals, solvents, fertilizers, pesticides, and plastics; the 16 percent not used for energy production is converted into these other materials. Petroleum is found in porous rock formations in the upper strata of some areas of the Earth's crust. There is also petroleum in oil sands (tar sands). Known oil reserves are typically estimated at around 190 km^3 (1.2 trillion (short scale) barrels) without oil sands, or 595 km^3 (3.74 trillion barrels) with oil sands. Consumption is currently around 84 million barrels (13.4×10^6 m^3) per day, or 4.9 km^3 per year. Which in turn yields a remaining oil supply of only about 120 years, if current demand remain static.

Reservoirs

Crude Oil Reservoirs

Three conditions must be present for oil reservoirs to form: a source rock rich in hydrocarbon material buried deep enough for subterranean heat to cook it into oil, a porous and permeable reservoir rock for it to accumulate in, and a cap rock (seal) or other mechanism that prevents it from escaping to the surface. Within these reservoirs, fluids will typically organize themselves like a three-layer cake with a layer of water below the oil layer and a layer of gas above it, although the different layers vary in size between reservoirs. Because most hydrocarbons are less dense than rock or water, they often migrate upward through adjacent rock layers until either reaching the surface or becoming trapped within porous rocks (known as reservoirs) by impermeable rocks above. However, the process is influenced by underground water flows, causing oil to migrate hundreds of kilometres horizontally or even short distances downward before becoming trapped in a reservoir. When hydrocarbons are concentrated in a trap, an oil field forms, from which the liquid can be extracted by drilling and pumping.

The reactions that produce oil and natural gas are often modelled as first order breakdown reactions, where hydrocarbons are broken down to oil and natural gas by a set of parallel reactions, and oil eventually breaks down to natural gas by another set of reactions. The latter set is regularly used in petrochemical plants and oil refineries.

Wells are drilled into oil reservoirs to extract the crude oil. "Natural lift" production methods that rely on the natural reservoir pressure to force the oil to the surface are usually sufficient for a while after reservoirs are first tapped. In some reservoirs, such as in the Middle East, the natural pressure is sufficient over a long time. The natural pressure in most reservoirs, however, eventually dissipates. Then the oil must be extracted using "artificial lift" means. Over time, these "primary" methods become less effective and "secondary" production methods may be used. A common secondary method is "waterflood" or injection of water into the reservoir to increase pressure and force the oil to the drilled shaft or "wellbore." Eventually "tertiary" or "enhanced" oil recovery methods may be used to increase the oil's flow characteristics by injecting steam, carbon dioxide and other gases or chemicals into the reservoir. In the United States, primary production methods account for less than 40 percent of the oil produced on a daily basis, secondary methods account for about half, and tertiary recovery the remaining 10 percent. Extracting oil (or "bitumen") from oil/tar sand and oil shale deposits requires mining the sand or shale and heating it in a vessel or retort, or using "in-situ" methods of injecting heated liquids into the deposit and then pumping out the oil-saturated liquid.

Unconventional Oil Reservoirs

Oil-eating bacteria biodegrade oil that has escaped to the surface. Oil sands are reservoirs of partially biodegraded oil still in the process of escaping and being biodegraded, but they contain so much migrating oil that, although most of it has escaped, vast amounts are still present—more than can be found in conventional oil reservoirs. The lighter fractions of the crude oil are destroyed first, resulting in reservoirs containing an extremely heavy form of crude oil, called crude bitumen in Canada, or extra-heavy crude oil in Venezuela. These two countries have the world's largest deposits of oil sands.

On the other hand, oil shales are source rocks that have not been exposed to heat or pressure long enough to convert their trapped hydrocarbons into crude oil. Technically speaking, oil shales are not always shales and do not contain oil, but are fined-grain sedimentary rocks containing an insoluble organic solid called kerogen. The kerogen in the rock can be converted into crude oil using heat and pressure to simulate natural processes. The method has been known for centuries and was patented in 1694 under British Crown Patent No. 330 covering, "A way to extract and make great quantities of pitch, tar, and oil out of a sort of stone." Although oil shales are found in many countries, the United States has the world's largest deposits.

Classification

The petroleum industry generally classifies crude oil by the geographic location it is produced in (e.g. West Texas Intermediate, Brent, or Oman), its API gravity (an oil industry measure of density), and its sulfur content. Crude oil may be considered *light* if it has low density or *heavy* if it has high density; and it may be referred to as sweet if it contains relatively little sulfur or *sour* if it contains substantial amounts of sulfur.

The geographic location is important because it affects transportation costs to the refinery. *Light* crude oil is more desirable than *heavy* oil since it produces a higher yield of petrol, while *sweet* oil commands a higher price than *sour* oil because it has fewer environmental problems and requires less refining to meet sulfur standards imposed on fuels in consuming countries. Each crude oil has unique molecular characteristics which are understood by the use of crude oil assay analysis in petroleum laboratories.

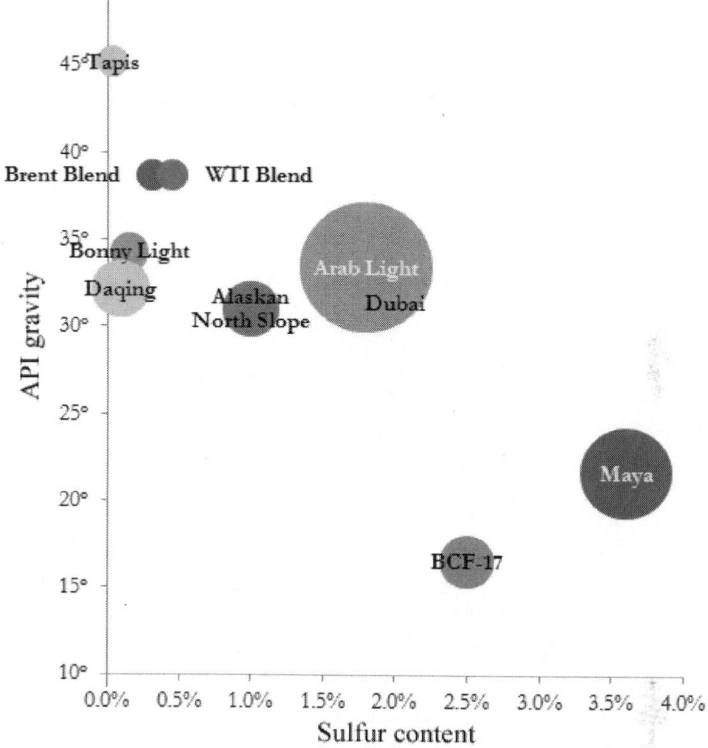

Figure: *Some marker crudes with their sulfur content (horizontal) and API gravity (vertical) and relative production quantity.*

Barrels from an area in which the crude oil's molecular characteristics have been determined and the oil has been classified are used as pricing references throughout the world. Some of the common reference crudes are:

- West Texas Intermediate (WTI), a very high-quality, sweet, light oil delivered at Cushing, Oklahoma for North American oil
- Brent Blend, comprising 15 oils from fields in the Brent and Ninian systems in the East Shetland Basin of the North Sea. The oil is landed at Sullom Voe terminal in Shetland. Oil production from Europe, Africa and Middle Eastern oil flowing West tends to be priced off this oil, which forms a benchmark
- Dubai-Oman, used as benchmark for Middle East sour crude oil flowing to the Asia-Pacific region

- Tapis (from Malaysia, used as a reference for light Far East oil)
- Minas (from Indonesia, used as a reference for heavy Far East oil)
- The OPEC Reference Basket, a weighted average of oil blends from various OPEC (The Organization of the Petroleum Exporting Countries) countries
- Midway Sunset Heavy, by which heavy oil in California is priced

There are declining amounts of these benchmark oils being produced each year, so other oils are more commonly what is actually delivered. While the reference price may be for West Texas Intermediate delivered at Cushing, the actual oil being traded may be a discounted Canadian heavy oil delivered at Hardisty, Alberta, and for a Brent Blend delivered at Shetland, it may be a Russian Export Blend delivered at the port of Primorsk.

Petroleum Industry

The petroleum industry is involved in the global processes of exploration, extraction, refining, transporting (often with oil tankers and pipelines), and marketing petroleum products. The largest volume products of the industry are fuel oil and petrol. Petroleum is also the raw material for many chemical products, including pharmaceuticals, solvents, fertilizers, pesticides, and plastics. The industry is usually divided into three major components: upstream, midstream and downstream. Midstream operations are usually included in the downstream category.

Petroleum is vital to many industries, and is of importance to the maintenance of industrialized civilization itself, and thus is a critical concern to many nations. Oil accounts for a large percentage of the world's energy consumption, ranging from a low of 32 percent for Europe and Asia, up to a high of 53 percent for the Middle East, South and Central America (44%), Africa (41%), and North America (40%). The world at large consumes 30 billion barrels (4.8 km^3) of oil per year, and the top oil consumers largely consist of developed nations. In fact, 24 percent of the oil consumed in 2004 went to the United States alone, though by 2007 this had dropped to 21 percent of world oil consumed.

In the US, in the states of Arizona, California, Hawaii, Nevada, Oregon and Washington, the Western States Petroleum Association (WSPA) represents companies responsible for producing, distributing, refining, transporting and marketing petroleum. This non-profit trade association was founded in 1907, and is the oldest petroleum trade association in the United States.

Shipping

In the 1950s, shipping costs made up 33 percent of the price of oil transported from the Persian Gulf to USA, but due to the development of supertankers in the 1970s, the cost of shipping dropped to only 5 percent of the price of Persian oil in USA. Due to the increase of the value of the crude oil during the last 30 years, the share of the shipping cost on the final cost of the delivered commodity was less than 3% in 2010. For example, in 2010 the shipping cost from the Persian Gulf to the USA was in the range of 20 $/t and the cost of the delivered crude oil around 800 $/t.

Fuels

The most common distillation fractions of petroleum are fuels. Fuels include (by increasing boiling temperature range):

Common fractions of petroleum as fuels

Fraction	*Boiling range °C*
Liquefied petroleum gas (LPG)	“40
Butane	“12 to “1
Petrol	“1 to 110
Jet fuel	150 to 205
Kerosene	205 to 260
Fuel oil	205 to 290
Diesel fuel	260 to 315

Petroleum classification according to chemical composition.

Class of petroleum	*Composition of 250°--300 °C fraction, wt. %*				
	Par.	Napth	Arom.	Wax	Asph.
Paraffinic	46—61	22–32	12–25	1.5--10	0—6 --6
Paraffinic-naphtenic	42–45	38–39	16–20	1--6	--6
Naphthenic	15–26	61–76	8--13	trace	0—6 --6`
Paraffinic-naphtenic-aromatic	27–35	36–47	26–33	0.5--1	0--10
Aromatic	0--8	57—78	20—25	0—0.5	0--20

Other Derivatives

Certain types of resultant hydrocarbons may be mixed with other non-hydrocarbons, to create other end products:

- Alkenes (olefins), which can be manufactured into plastics or other compounds

- Lubricants (produces light machine oils, motor oils, and greases, adding viscosity stabilizers as required)
- Wax, used in the packaging of frozen foods, among others
- Sulfur or Sulfuric acid. These are useful industrial materials. Sulfuric acid is usually prepared as the acid precursor oleum, a byproduct of sulfur removal from fuels.
- Bulk tar
- Asphalt
- Petroleum coke, used in speciality carbon products or as solid fuel
- Paraffin wax
- Aromatic petrochemicals to be used as precursors in other chemical production

Agriculture

Since the 1940s, agricultural productivity has increased dramatically, due largely to the increased use of energy-intensive mechanization, fertilizers and pesticides.

Ocean Resources

The ocean is one of Earth's most valuable natural resources. It provides food in the form of fish and shellfish—about 200 billion pounds are caught each year. It's used for transportation—both travel and shipping. It provides a treasured source of recreation for humans. It is mined for minerals (salt, sand, gravel, and some manganese, copper, nickel, iron, and cobalt can be found in the deep sea) and drilled for crude oil.

The ocean plays a critical role in removing carbon from the atmosphere and providing oxygen. It regulates Earth's climate. The ocean is an increasingly important source of biomedical organisms with enormous potential for fighting disease. These are just a few examples of the importance of the ocean to life on land. Explore them in greater detail to understand why we must keep the ocean healthy for future generations.

Fishing Facts

The oceans have been fished for thousands of years and are an integral part of human society. Fish have been important to the world economy for all of these years, starting with the Viking trade of cod and then continuing with fisheries like those found in Lofoten, Europe, Italy, Portugal, Spain and India. Fisheries of today provide about 16%

of the total world's protein with higher percentages occurring in developing nations. Fisheries are still enormously important to the economy and wellbeing of communities.

The word fisheries refers to all of the fishing activities in the ocean, whether they are to obtain fish for the commercial fishing industry, for recreation or to obtain ornamental fish or fish oil. Fishing activities resulting in fish not used for consumption are called industrial fisheries. Fisheries are usually designated to certain ecoregions like the salmon fishery in Alaska, the Eastern Pacific tuna fishery or the Lofoten island cod fishery. Due to the relative abundance of fish on the continental shelf, fisheries are usually marine and not freshwater.

Although a world total of 86 million tons of fish were captured in 2000, China's fisheries were the most productive, capturing a whopping one third of the total. Other countries producing the most fish were Peru, Japan, the United States, Chile, Indonesia, Russia, India, Thailand, Norway and Iceland- with Peru being the most and Iceland being the least. The number of fish caught varies with the years, but appears to have leveled off at around 88 million tons per year possibly due to overfishing, economics and management practices.

Fish are caught in a variety of ways, including one-man casting nets, huge trawlers, seining, driftnetting, handlining, longlining, gillnetting and diving. The most common species making up the global fisheries are herring, cod, anchovy, flounder, tuna, shrimp, mullet, squid, crab, salmon, lobster, scallops and oyster. Mollusks and crustaceans are also widely sought. The fish that are caught are not always used for food. In fact, about 40% of fish are used for other purposes such as fishmeal to feed fish grown in captivity. For example cod, is used for consumption, but is also frozen for later use. Atlantic herring is used for canning, fishmeal and fish oil. The Atlantic menhaden is used for fishmeal and fish oil and Alaska pollock is consumed, but also used for fish paste to simulate crab. The Pacific cod has recently been used as a substitute for Atlantic cod which has been overfished.

The amount of fish available in the oceans is an ever-changing number due to the effects of both natural causes and human developments. It will be necessary to manage ocean fisheries in the coming years to make sure the number of fish caught never makes it to zero. A lack of fish greatly impacts the economy of communities dependent on the resource, as can be seen in Japan, eastern Canada, New England, Indonesia and Alaska. The anchovy fisheries off the coast of western South America have already collapsed and with numbers dropping violently from 20 million tons to 4 million tons—

they may never fully recover. Other collapses include the California sardine industry, the Alaskan king crab industry and the Canadian northern cod industry. In Massachusetts alone, the cod, haddock and yellowtail flounder industries collapsed, causing an economic disaster for the area.

Due to the importance of fishing to the worldwide economy and the need for humans to understand human impacts on the environment, the academic division of fisheries science was developed. Fisheries science includes all aspects of marine biology, in addition to economics and management skills and information. Marine conservation issues like overfishing, sustainable fisheries and management of fisheries are also examined through fisheries science.

In order for there to be plenty of fish in the years ahead, fisheries will have to develop sustainable fisheries and some will have to close. Due to the constant increase in the human population, the oceans have been overfished with a resulting decline of fish crucial to the economy and communities of the world. The control of the world's fisheries is a controversial subject, as they cannot produce enough to satisfy the demand, especially when there aren't enough fish left to breed in healthy ecosystems. Scientists are often in the role of fisheries managers and must regulate the amount of fishing in the oceans, a position not popular with those who have to make a living fishing ever decreasing populations.

The two main questions facing fisheries management are:

1. What is the carrying capacity of the ocean? How many fish are there and how many of which type of fish should be caught to make fisheries sustainable?
2. How should fisheries resources be divided among people?

Fish populate the ocean in patches instead of being spread out throughout the enormous expanse. The photic zone is only 10-30 m deep near the coastline, a place where phytoplankton have enough solar energy to grow in abundance and fish have enough to eat. Most commercial fishing takes place in these coastal waters, as well as estuaries and the slope of the Continental Shelf. High nutrient contents from upwelling, runoff, the regeneration of nutrients and other ecological processes supply fish in these areas with the necessary requirements for life. The blue colour of the water near the coastlines is the result of chlorophyll contained in aquatic plant life.

Most fish are only found in very specific habitats. Shrimp are fished in river deltas that bring large amounts of freshwater into the ocean.

The areas of highest productivity known as banks are actually where the Continental Shelf extends outward towards the ocean. These include the Georges Bank near Cape Cod, the Grand Banks near Newfoundland and Browns Bank. Areas where the ocean is very shallow also contain many fish and include the middle and southern regions of the North Sea. Coastal upwelling areas can be found off of southwest Africa and off South America's western coast. In the open ocean, tuna and other mobile species like yellowfin can be found in large amounts.

The question of how many fish there are in the ocean is a complicated one but can be simplified using populations of fish instead of individuals. The word "cohort" refers to the year the fish was born and is used to gather population statistics. Cohorts start off as eggs with an extremely high rate of mortality, which declines as the fish gets older. Juvenile fish close to the age where they can be fished are called "recruits". Cohort mortality is tied in with the species of fish due to variances in natural mortality. The biomass of a particular cohort is greatest when fish are rapidly growing and decreases as the fish get older and start to die.

Scientists use theories and models to help determine the number and size of fish populations in the ocean. Production theory is the theory that production will be highest when the number of fish does not overwhelm the environment and there are not too few for genetic diversity of populations. The maximum sustainable yield is produced when the population is of intermediate size. Yield-per-recruit theory is the quest to determine the optimum age for harvesting fish. The theory of recruitment and stock allows scientists to make a guess about the optimum population size to encourage a larger population of recruits. All of the above theories must be flexible enough to allow natural fluctuations in the fish population to occur and still gather significant data; however, the theories are limited when taking into account the effect of humans on the environment and misinformation could result in overfishing of the ocean's resources.

Other factors that must be taken into account are the ecological requirements of individual fish species like predation and nutrition and why fish will often migrate to different areas. Water temperatures also influence the behaviour of ecosystems, causing an increase in metabolism and predation or a sort of hibernation. Even the amount of turbulence in the water can affect predator-prey relationships, with more meetings between the two when waters are stirred up. Global warming could have a huge economic impact on the fisheries when fish stocks are forced to move to waters with more tolerable temperatures.

In many countries, commercial fishing has found more temporarily economical ways of catching fish, including gill nets, purse seines, and drift nets. Although fish are trapped efficiently in one day using these fishing practices, the number of fish that are wasted this way has reached 27 million tons per year, not to mention the crucial habitats destroyed that are essential for the regeneration of fish stocks. In addition, marine mammals and birds are also caught in these nets. The wasted fish and marine life is referred to as bycatch, an unfortunate side-effect of unsustainable fishing practices that can turn the ecosystem upside-down and leave huge amounts of dead matter in the water. Other human activities like trawling and dredging of the ocean floor have bulldozed over entire underwater habitats. The oyster habitat has been completely destroyed in many areas from the use of the oyster patent tong and sediment buildup draining from farm runoff.

Shipping

The word "shipping" refers to the activity of moving cargo with ships in between seaports. Wind-powered ships exist, but more often ships are powered by steam turbine plants or diesel engines. Naval ships are usually responsible for transporting most of trade from one country to another and are called merchant navies. The various types of ships include container ships, tankers, crude oil ships, product ships, chemical ships, bulk carriers, cable layers, general cargo ships, offshore supply vessels, dynamically-positioned ships, ferries, gas and car carriers, tugboats, barges and dredgers.

In theory, shipping can have a low impact on the environment. It is safe and profitable for economies around the world. However, serious problems occur with the shipping of oil, dumping of waste water into the ocean, chemical accidents at sea, and the inevitable air and water pollution occuring when modern day engines are used. Ships release air pollutants in the form of sulphur dioxide, nitrogen oxides, carbon dioxide, hydrocarbons and carbon monoxide. Chemicals dumped in the ocean from ships include chemicals from the ship itself, cleaning chemicals for machine parts, and cleaning supplies for living quarters. Large amounts of chemicals are often spilled into the ocean and sewage is not always treated properly or treated at all. Alien species riding in the ballast water of ships arrive in great numbers to crash native ecosystems and garbage is dumped over the side of many vessels. Dangerous industrial waste and harmful substances like halogenated hydrocarbons, water treatment chemicals, and antifouling paints are also dumped frequently. Ships and other watercraft with engines disturb the natural environment with loud noises, large waves, frequently striking and killing animals like manatees and dolphins.

Tourism

Tourism is the fastest growing division of the world economy and is responsible for more than 200 million jobs all over the world. In the US alone, tourism resulted in an economic gain of 478 billion dollars. With 700 million people travelling to another country in the year 2000, tourism is in the top five economic contributors to 83% of all countries and the most important economy for 38% of countries. The tourism industry is based on natural resources present in each country and usually negatively affect ecosystems because it is often left unmanaged. However, sustainable tourism can actually promote conservation of the environment.

The negative effects of tourism originate from the development of coastal habitats and the annihilation of entire ecosystems like mangroves, coral reefs, wetlands and estuaries. Garbage and sewage generated by visitors can add to the already existing solid waste and garbage disposal issues present in many communities. Often visitors produce more waste than locals, and much of it ends up as untreated sewage dumped in the ocean. The ecosystem must cope with eutrophication, or the loss of oxygen in the water due to excessive algal bloom, as well as disease epidemics. Sewage can be used as reclaimed water to treat lawns so that fertilizers and pesticides do not seep into the ocean.

Other problems with tourism include the overexploitation of local seafood, the destruction of local habitats through careless scuba diving or snorkeling and the dropping of anchors on underwater features. Ecotourism and cultural tourism are a new trend that favours low impact tourism and fosters a respect for local cultures and ecosystems.

Mining

Humans began to mine the ocean floor for diamonds, gold, silver, metal ores like manganese nodules and gravel mines in the 1950's when the company Tidal Diamonds was established by Sam Collins. Diamonds are found in greater number and quality in the ocean than on land, but are much harder to mine. When diamonds are mined, the ocean floor is dredged to bring it up to the boat and sift through the sediment for valuable gems. The process is difficult as sediment is not easy to bring up to the surface, but will probably become a huge industry once technology evolves to solve the logistical problem.

Metal compounds, gravels, sands and gas hydrates are also mined in the ocean. Mining of manganese nodules containing nickel, copper and cobalt began in the 1960's and soon after it was discovered that

Papua New Guinea was one of the few places where nodules were located in shallow waters rather than deep waters. Although manganese nodules could be found in shallow waters in significant quantities, the expense of bringing the ore up to the surface proved to be expensive. Sands and gravels are often mined for in the United States and are used to protect beaches and reduce the effects of erosion.

Mining the ocean can be devastating to the natural ecosystems. Dredging of any kind pulls up the ocean floor resulting in widespread destruction of marine animal habitats, as well as wiping out vast numbers of fishes and invertebrates. When the ocean floor is mined, a cloud of sediment rises up in the water, interfering with photosynthetic processes of phytoplankton and other marine life, in addition to introducing previously benign heavy metals into the food chain. As minerals found on land are exploited and used up, mining of the ocean floor will increase.

Climate Buffer

The ocean is an integral component of the world's climate due to its capacity to collect, drive and mix water, heat, and carbon dioxide. The ocean can hold and circulate more water, heat and carbon dioxide than the atmosphere although the components of the Earth's climate are constantly exchanged. Because the ocean can store so much heat, seasons occur later than they would and air above the ocean is warmed. Heat energy stored in the ocean in one season will affect the climate almost an entire season later. The ocean and the atmosphere work together to form complex weather phenomena like the North Atlantic Oscillation and El Niño. The many chemical cycles occurring between the ocean and the atmosphere also influence the climate by controlling the amount of radiation released into ecosystems and our environment.

The atmosphere directly above the ocean does not absorb much heat by itself, so in order for it to warm up, the temperature of the ocean has to rise first. The two other ways for the atmosphere to warm near the ocean are by reflection of light off of the surface of the ocean or by the evaporation of water from the ocean surface. The temperature of the ocean controls the climate in the lower part of the atmosphere, so for most areas of the Earth the ocean temperature is responsible for the air temperature. The main forms of climate buffering by the ocean are by the transport of heat through ocean currents travelling across huge basins. Areas like the tropics end up being cooled and higher latitudes are warmed by this effect.

Air temperatures worldwide are regulated by the circulation of heat by the oceans. The ocean stores heat in the upper two metres of the photic zone. This is possible because seawater has a very high density and specific heat and can store vast quantities of energy in the form of heat. The ocean can then buffer changes in temperature by storing heat and releasing heat. Evaporation cools ocean water which cools the atmosphere. It is most noticeable near the equator and the effect decreases closer to the poles.

Oxygen Production

Gases in the atmosphere like carbon, nitrogen, sulfur and oxygen are dissolved through the water cycle. The gases that are now crucial to all ecosystems and biological processes originally came from the inside layers of the earth during the period when the earth was first formed. The rate of flow for oxygen as well as other gases is controlled by biological processes, especially metabolism of organisms like prokaryotes and bacteria. Prokaryotes have been around since the beginning of the Earth, have evolved to be able to use chemical energy to create organic matter and are capable of both reducing and oxidizing inorganic compounds. Bacteria that can reduce inorganic compounds are anaerobic and those that oxidize inorganic compounds are aerobic. Aerobic bacteria release oxygen as a by-product of photosynthesis.

Approximately two billion years ago, aerobic bacteria began producing oxygen which gradually filled up all of the oxygen reservoirs in the environment. Once these "sinks" were filled, molecular oxygen began to build in the atmosphere, creating an environment favourable for other life to inhabit the Earth. Sinks included reduced iron ions and hydrogen sulfide gas. Evidence of this process can be found in the banded iron formations created when iron minerals were precipitated. The oxygen started to fill the atmosphere up and new bacteria evolved that could use oxygen to oxidize both inorganic and organic compounds. Bacteria that were accustomed to an oxygen-poor atmosphere only survived in anaerobic environments like sewage, swamps, and in the sediments of both marine and freshwater areas.

Phytoplankton account for possibly 90% of the world's oxygen production because water covers about 70% of the Earth and phytoplankton are abundant in the photic zone of the surface layers. Some of the oxygen produced by phytoplankton is absorbed by the ocean, but most flows into the atmosphere where it becomes available for oxygen dependent life forms.

Chapter 5

Natural Resource Management

Natural resource management refers to the management of natural resources such as land, water, soil, plants and animals, with a particular focus on how management affects the quality of life for both present and future generations (stewardship).

Natural resource management deals with managing the way in which people and natural landscapes interact. It brings together land use planning, water management, biodiversity conservation, and the future sustainability of industries like agriculture, mining, tourism, fisheries and forestry. It recognises that people and their livelihoods rely on the health and productivity of our landscapes, and their actions as stewards of the land play a critical role in maintaining this health and productivity.

Natural resource management specifically focuses on a scientific and technical understanding of resources and ecology and the life-supporting capacity of those resources. Environmental management is also similar to natural resource management. In academic contexts, the sociology of natural resources is closely related to, but distinct from, natural resource management.

History

The emphasis on sustainability can be traced back to early attempts to understand the ecological nature of North American rangelands in the late 19th century, and the resource conservation movement of the same time. This type of analysis coalesced in the 20th century with recognition that preservationist conservation strategies had not been effective in halting the decline of natural resources. A more integrated approach was implemented recognising the intertwined social, cultural, economic and political aspects of resource management. A more holistic, national and even global form evolved, from the Brundtland Commission and the advocacy of sustainable development.

In 2005 the government of New South Wales, established a *Standard for Quality Natural Resource Management*, to improve the consistency of practice, based on an adaptive management approach.

In the United States, the most active areas of natural resource management are wildlife management often associated with ecotourism and rangeland (pastures) management. In Australia, water sharing, such as the Murray Darling Basin Plan and catchment management are also significant.

Ownership Regimes

Natural resource management approaches can be categorised according to the kind and right of stakeholders, natural resources:

- State property
- Private property
- Common property
- Non-property (open access)
- Hybrid

State Property Regime

Ownership and control over the use of resources is in hands of the state. Individuals or groups may be able to make use of the resources, but only at the permission of the state. National forest, National parks and military reservations are some US examples..

Private Property Regime

Any property owned by a defined individual or corporate entity. Both the benefit and duties to the resources fall to the owner(s). Private land is the most common example.

Common Property Regimes

It is a private property of a group. The group may vary in size, nature and internal structure e.g. indigenous tribe, neighbours of village. Some examples of common property are community forests and water resources.

Non-property Regimes (Open Access)

There is no definite owner of these properties. Each potential user has equal ability to use it as they wish. These areas are the most exploited. It is said that "Everybody's property is nobody's property".

An example is a lake fishery. This ownership regime is often linked to the tragedy of the commons.

Hybrid Regimes

Many ownership regimes governing natural resources will contain parts of more than one of the regimes described above, so natural resource managers need to consider the impact of hybrid regimes. An example of such a hybrid is native vegetation management in NSW, Australia, where legislation recognises a public interest in the preservation of native vegetation, but where most native vegetation exists on private land.

Stakeholder analysis

Stakeholder analysis originated from business management practices and has been incorporated into natural resource management in ever growing popularity. Stakeholder analysis in the context of natural resource management identifies distinctive interest groups affected in the utilisation and conservation of natural resources.

There is no definitive definition of a stakeholder as illustrated in the table below. Especially in natural resource management as it is difficult to determine who has a stake and this will differ according to each potential stakeholder.

Different Approaches to who is a Stakeholder

Source	*Who is a stakeholder*	*Kind of research*
Freeman.	"can affect or is affected by the achievement of the organization's objectives"	Business Management
Bowie	"without whose support the organization would cease to exist"	Business Management
Clarkson	"...persons or groups that have, or claim, ownership, rights, or interests in a corporation and its activities, past, present, or future."	Business Management
Grimble and Wellard	"...any group of people, organized or unorganized, who share a common interest or stake in a particular issue or system..."	Natural resource management
Gass et al.	"...any individual, group and institution who would potentially be affected, whether positively or negatively, by a specified event, process or change."	Natural resource management
Buanes et al	"...any group or individual who may directly or indirectly affect—or be affected—...planning to be at least potential stakeholders."	Natural resource management
Brugha and Varvasovszky	"...actors who have an interest in the issue under consideration, who are affected by the issue, or who—because of their position—have or could have an active or passive influence on the decision making and implementation process."	Health policy
ODA	"... persons, groups or institutions with interests in a project or programme."	Development

Therefore it is dependent upon the circumstances of the stakeholders involved with natural resource as to which definition and subsequent theory is utilised. Billgrena and Holme identified the aims of stakeholder analysis in natural resource management:

- Identify and categorise the stakeholders that may have influence
- Develop an understanding of why changes occur
- Establish who can make changes happen
- How to best manage natural resources

This gives transparency and clarity to policy making allowing stakeholders to recognise conflicts of interest and facilitate resolutions. There are numerous stakeholder theories such as Mitchell et al. however Grimble created a framework of stages for a Stakeholder Analysis in natural resource management. Grimble designed this framework to ensure that the analysis is specific to the essential aspects of natural resource management.

Stages in Stakeholder Analysis

1. Clarify objectives of the analysis
2. Place issues in a systems context
3. Identify decision-makers and stakeholders
4. Investigate stakeholder interests and agendas
5. Investigate patterns of inter-action and dependence (e.g. conflicts and compatibilities, trade-offs and synergies)

Application

Grimble and Wellard established that Stakeholder analysis in natural resource management is most relevant where issued can be characterised as;

- Cross-cutting systems and stakeholder interests
- Multiple uses and users of the resource.
- Market failure
- Subtractability and temporal trade-offs
- Unclear or open-access property rights
- Untraded products and services
- Poverty and under-representation

Case Studies

In the case of the Bwindi Impenetrable National Park, a comprehensive stakeholder analysis would have been relevant and the

Batwa people would have potentially been acknowledged as stakeholders preventing the loss of people's livelihoods and loss of life.

Nepal, Indonesia and Koreas' community forestry are successful examples of how stakeholder analysis can be incorporated into the management of natural resources. This allowed the stakeholders to identify their needs and level of involvement with the forests.

Criticisms

- Natural resource management stakeholder analysis tends to include too many stakeholders which can create problems in of its self as suggested by Clarkson. "Stakeholder theory should not be used to weave a basket big enough to hold the world's misery."
- Starik proposed that nature needs to be represented as stakeholder. However this has been rejected by many scholars as it would be difficult to find appropriate representation and this representation could also be disputed by other stakeholders causing further issues.
- Stakeholder analysis can be used exploited and abused in order to marginalise other stakeholders.
- Identifying the relevant stakeholders for participatory processes is complex as certain stakeholder groups may have been excluded from previous decisions.
- On-going conflicts and lack of trust between stakeholders can prevent compromise and resolutions.

Alternatives/ Complementary forms of analysis:

- Social Network Analysis
- Common Pool Resource

Management Approaches

Natural resource management issues are inherently complex as they involve the ecological cycles, hydrological cycles, climate, animals, plants and geography etc. All these are dynamic and inter-related. A change in one of them may have far reaching and/or long term impacts which may even be irreversible.

In addition to the natural systems, natural resource management also has to manage various stakeholders and their interests, policies, politics, geographical boundaries, economic implications and the list goes on. It is very difficult to satisfy all aspects at the same time. This results in conflicting situations.

After the United Nations Conference for the Environment and Development (UNCED) held in Rio de Janeiro in 1992, most nations subscribed to new principles for the integrated management of land, water, and forests. Although program names vary from nation to nation, all express similar aims.

The various approaches applied to natural resource management include:

- Top-down (command and control)
- Community-based natural resource management
- Adaptive management
- Precautionary approach
- Integrated natural resource management)

Community-based Natural Resource Management

The community-based natural resource management (CBNRM) approach combines conservation objectives with the generation of economic benefits for rural communities. The three key assumptions being that: locals are better placed to conserve natural resources, people will conserve a resource only if benefits exceed the costs of conservation, and people will conserve a resource that is linked directly to their quality of life. When a local people's quality of life is enhanced, their efforts and commitment to ensure the future well-being of the resource are also enhanced. Regional and community based natural resource management is also based on the principle of subsidiarity.

The United Nations advocates CBNRM in the Convention on Biodiversity and the Convention to Combat Desertification. Unless clearly defined, decentralised NRM can result an ambiguous socio-legal environment with local communities racing to exploit natural resources while they can e.g. forest communities in central Kalimantan (Indonesia).

A problem of CBNRM is the difficulty of reconciling and harmonising the objectives of socioeconomic development, biodiversity protection and sustainable resource utilisation. The concept and conflicting interests of CBNRM, show how the motives behind the participation are differentiated as either people-centred (active or participatory results that are truly empowering) or planner-centred (nominal and results in passive recipients). Understanding power relations is crucial to the success of community based NRM. Locals may be reluctant to challenge government recommendations for fear of losing promised benefits.

CBNRM is based particularly on advocacy by nongovernmental organizations working with local groups and communities, on the one hand, and national and transnational organizations, on the other, to build and extend new versions of environmental and social advocacy that link social justice and environmental management agendas with both direct and indirect benefits observed including a share of revenues, employment, diversification of livelihoods and increased pride and identity. CBNRM has raised new challenges, as concepts of community, territory, conservation, and indigenous are worked into politically varied plans and programs in disparate sites. Warner and Jones address strategies for effectively managing conflict in CBNRM.

The capacity of indigenous communities to conserve natural resources has been acknowledged by the Australian Government with the Caring for Country Program. Caring for our Country is an Australian Government initiative jointly administered by the Australian Government Department of Agriculture, Fisheries and Forestry and the Department of the Environment, Water, Heritage and the Arts. These Departments share responsibility for delivery of the Australian Government's environment and sustainable agriculture programs, which have traditionally been broadly referred to under the banner of 'natural resource management'.

These programs have been delivered regionally, through 56 State government bodies, successfully allowing regional communities to decide the natural resource priorities for their regions.

Governance is seen as a key consideration for delivering community-based or regional natural resource management. In the State of NSW, the 13 catchment management authorities (CMAs) are overseen by the Natural Resources Commission (NRC), responsible for undertaking audits of the effectiveness of regional natural resource management programs.

Adaptive Management

The primary methodological approach adopted by catchment management authorities (CMAs) for regional natural resource management in Australia is adaptive management.

This approach includes recognition that adaption occurs through a process of 'plan-do-review-act'. It also recognises seven key components that should be considered for quality natural resource management practice:

- Determination of scale
- Collection and use of knowledge

- Information management
- Monitoring and evaluation
- Risk management
- Community engagement
- Opportunities for collaboration.

Integrated Natural Resource Management

Integrated natural resource management (INRM) is a process of managing natural resources in a systematic way, which includes multiple aspects of natural resource use (biophysical, socio-political, and economic) meet production goals of producers and other direct users (e.g., food security, profitability, risk aversion) as well as goals of the wider community (e.g., poverty alleviation, welfare of future generations, environmental conservation).

It focuses on sustainability and at the same time tries to incorporate all possible stakeholders from the planning level itself, reducing possible future conflicts. The conceptual basis of INRM has evolved in recent years through the convergence of research in diverse areas such as sustainable land use, participatory planning, integrated watershed management, and adaptive management. INRM is being used extensively and been successful in regional and community based natural management.

Frameworks and Modelling

There are various frameworks and computer models developed to assist natural resource management.

Geographic Information Systems (GIS)

GIS is a powerful analytical tool as it is capable of overlaying datasets to identify links. A bush regeneration scheme can be informed by the overlay of rainfall, cleared land and erosion. In Australia, Metadata Directories such as NDAR provide data on Australian natural resources such as vegetation, fisheries, soils and water. These are limited by the potential for subjective input and data manipulation.

Natural Resources Management Audit Frameworks

The NSW Government in Australia has published an audit framework for natural resource management, to assist the establishment of a performance audit role in the governance of regional natural resource management. This audit framework builds from other established audit methodologies, including performance audit, environmental audit and internal audit. Audits undertaken using this framework have provided confidence to stakeholders, identified areas for improvement and described policy expectations for the general public.

The Australian Government has established a framework for auditing greenhouse emissions and energy reporting, which closely follows Australian Standards for Assurance Engagements.

The Australian Government is also currently preparing an audit framework for auditing water management, focussing on the implementation of the Murray Darling Basin Plan.

Other Elements

Biodiversity Conservation

The issue of biodiversity conservation is regarded as an important element in natural resource management. What is biodiversity? Biodiversity is a comprehensive concept, which is a description of the extent of natural diversity. Gaston and Spicer (p. 3) point out that biodiversity is "the variety of life" and relate to different kinds of "biodiversity organization". According to Gray (p. 154), the first widespread use of the definition of biodiversity, was put forward by the United Nations in 1992, involving different aspects of biological diversity.

Precautionary Biodiversity Management

The "threats" wreaking havoc on biodiversity include; habitat fragmentation, putting a strain on the already stretched biological resources; forest deterioration and deforestation; the invasion of "alien species" and "climate change"(p. 2). Since these threats have received increasing attention from environmentalists and the public, the precautionary management of biodiversity becomes an important part of natural resources management. According to Cooney, there are material measures to carry out precautionary management of biodiversity in natural resource management.

Concrete "Policy Tools"

Cooney claims that the policy making is dependent on "evidences", relating to "high standard of proof", the forbidding of special "activities" and "information and monitoring requirements". Before making the policy of precaution, categorical evidence is needed. When the potential menace of "activities" is regarded as a critical and "irreversible" endangerment, these "activities" should be forbidden. For example, since explosives and toxicants will have serious consequences to endanger human and natural environment, the South Africa Marine

Living Resources Act promulgated a series of policies on completely forbidding to "catch fish" by using explosives and toxicants.

Administration and Guidelines

According to Cooney, there are 4 methods to manage the precaution of biodiversity in natural resources management;

1. "Ecosystem based Management" including "more risk-averse and precautionary management", where "given prevailing uncertainty regarding ecosystem structure, function, and inter-specific interactions, precaution demands an ecosystem rather than single-species approach to management".
2. "Adaptive management" is "a management approach that expressly tackles the uncertainty and dynamism of complex systems".
3. "Environmental impact assessment" and exposure ratings decrease the "uncertainties" of precaution, even though it has deficiencies, and
4. "Protectionist approaches", which "most frequently links to" biodiversity conservation in natural resources management.

Land Management

In order to have a sustainable environment, understanding and using appropriate management strategies is important. In terms of understanding, Young emphasises some important points of land management:

- Comprehending the processes of nature including ecosystem, water, soils
- Using appropriate and adapting management systems in local situations
- Cooperation between scientists who have knowledge and resources and local people who have knowledge and skills

Dale et al. (2000) study has shown that there are five fundamental and helpful ecological principles for the land manager and people who need them. The ecological principles relate to time, place, species, disturbance and the landscape and they interact in many ways. It is suggested that land managers could follow these guidelines:

- Examine impacts of local decisions in a regional context, and the effects on natural resources.
- Plan for long-term change and unexpected events.
- Preserve rare landscape elements and associated species.

- Avoid land uses that deplete natural resources.
- Retain large contiguous or connected areas that contain critical habitats.
- Minimize the introduction and spread of non-native species.
- Avoid or compensate for the effects of development on ecological processes.
- Implement land-use and land-management practices that are compatible with the natural potential of the area.

Food

Food is any substance consumed to provide nutritional support for the body. It is usually of plant or animal origin, and contains essential nutrients, such as fats, proteins, vitamins, or minerals. The substance is ingested by an organism and assimilated by the organism's cells to provide energy, maintain life, or stimulate growth.

Historically, people secured food through two methods: hunting and gathering, and agriculture. Today, most of the food energy required by the ever increasing population of the world is supplied by the food industry. Food safety and food security are monitored by agencies like the International Association for Food Protection, World Resources Institute, World Food Programme, Food and Agriculture Organization, and International Food Information Council. They address issues such as sustainability, biological diversity, climate change, nutritional economics, population growth, water supply, and access to food.

The right to food is a human right derived from the International Covenant on Economic, Social and Cultural Rights (ICESCR), recognizing the "right to an adequate standard of living, including adequate food," as well as the "fundamental right to be free from hunger."

Food Sources

Most food has its origin in plants. Some food is obtained directly from plants; but even animals that are used as food sources are raised by feeding them food derived from plants. Cereal grain is a staple food that provides more food energy worldwide than any other type of crop. Maize, wheat, and rice – in all of their varieties – account for 87% of all grain production worldwide. Most of the grain that is produced worldwide is fed to livestock.

Some foods not from animal or plant sources include various edible fungi, especially mushrooms. Fungi and ambient bacteria are used in the preparation of fermented and pickled foods like leavened bread, alcoholic drinks, cheese, pickles, kombucha, and yogurt. Another example is blue-green algae such as Spirulina. Inorganic substances such as salt, baking soda and cream of tartar are used to preserve or chemically alter an ingredient.

Plants

Many plants or plant parts are eaten as food. There are around 2,000 plant species which are cultivated for food, and many have several distinct cultivars.

Seeds of plants are a good source of food for animals, including humans, because they contain the nutrients necessary for the plant's initial growth, including many healthful fats, such as Omega fats. In fact, the majority of food consumed by human beings are seed-based foods. Edible seeds include cereals (maize, wheat, rice, et cetera), legumes (beans, peas, lentils, et cetera), and nuts. Oilseeds are often pressed to produce rich oils - sunflower, flaxseed, rapeseed (including canola oil), sesame, et cetera.

Seeds are typically high in unsaturated fats and, in moderation, are considered a health food, although not all seeds are edible. Large seeds, such as those from a lemon, pose a choking hazard, while seeds from cherries and apples contain cyanide which could be poisonous only if consumed in large volumes.

Fruits are the ripened ovaries of plants, including the seeds within. Many plants and animals have coevolved such that the fruits of the former are an attractive food source to the latter, because animals that eat the fruits may excrete the seeds some distance away. Fruits, therefore, make up a significant part of the diets of most cultures. Some botanical fruits, such as tomatoes, pumpkins, and eggplants, are eaten as vegetables.

Vegetables are a second type of plant matter that is commonly eaten as food. These include root vegetables (potatoes and carrots), bulbs (onion family), leaf vegetables (spinach and lettuce), stem vegetables (bamboo shoots and asparagus), and inflorescence vegetables (globe artichokes and broccoli and other vegetables such as cabbage or cauliflower).

Animals

Animals are used as food either directly or indirectly by the products they produce. Meat is an example of a direct product taken from an animal, which comes from muscle systems or from organs. Food products produced by animals include milk produced by mammary glands, which in many cultures is drunk or processed into dairy products (cheese, butter, etc.). In addition, birds and other animals lay eggs, which are often eaten, and bees produce honey, a reduced nectar from flowers, which is a popular sweetener in many cultures.

Some cultures consume blood, sometimes in the form of blood sausage, as a thickener for sauces, or in a cured, salted form for times of food scarcity, and others use blood in stews such as jugged hare.

Some cultures and people do not consume meat or animal food products for cultural, dietary, health, ethical, or ideological reasons. Vegetarians choose to forgo food from animal sources to varying degrees. Vegans do not consume any foods that are or contain ingredients from an animal source.

Production

Most food has always been obtained through agriculture. With increasing concern over both the methods and products of modern industrial agriculture, there has been a growing trend toward sustainable agricultural practices.

This approach, partly fuelled by consumer demand, encourages biodiversity, local self-reliance and organic farming methods. Major influences on food production include international organizations (e.g. the World Trade Organization and Common Agricultural Policy), national government policy (or law), and war.

In popular culture, the mass production of food, specifically meats such as chicken and beef, has come under fire from various documentaries, most recently Food, Inc, documenting the mass slaughter and poor treatment of animals, often for easier revenues from large corporations. Along with a current trend towards environmentalism, people in Western culture have had an increasing trend towards the use of herbal supplements, foods for a specific group of person (such as dieters, women, or athletes), functional foods (fortified foods, such as omega-3 eggs), and a more ethnically diverse diet.

Several organisations have begun calling for a new kind of agriculture in which agroecosystems provide food but also support vital ecosystem services so that soil fertility and biodiversity are maintained rather than compromised.

According to the International Water Management Institute and UNEP, well-managed agroecosystems not only provide food, fibre and animal products, they also provide services such as flood mitigation, groundwater recharge, erosion control and habitats for plants, birds fish and other animals.

Food Manufacturing

Packaged foods are manufactured outside the home for purchase. This can be as simple as a butcher preparing meat, or as complex as a modern international food industry. Early food processing techniques were limited by available food preservation, packaging, and transportation. This mainly involved salting, curing, curdling, drying, pickling, fermenting, and smoking. Food manufacturing arose during the industrial revolution in the 19th century. This development took advantage of new mass markets and emerging new technology, such as milling, preservation, packaging and labelling, and transportation. It brought the advantages of pre-prepared time-saving food to the bulk of ordinary people who did not employ domestic servants.

At the start of the 21st century, a two-tier structure has arisen, with a few international food processing giants controlling a wide range of well-known food brands. There also exists a wide array of small local or national food processing companies. Advanced technologies have also come to change food manufacture. Computer-based control systems, sophisticated processing and packaging methods, and logistics and distribution advances can enhance product quality, improve food safety, and reduce costs.

Commercial Trade

International food Imports and Exports

The World Bank reported that the European Union was the top food importer in 2005, followed at a distance by the USA and Japan. Britain's need for food was especially well illustrated in World War II. Despite the implementation of food rationing, Britain remained dependent on food imports and the result was a long term engagement in the Battle of the Atlantic.

Food is traded and marketed on a global basis. The variety and availability of food is no longer restricted by the diversity of locally grown food or the limitations of the local growing season. Between 1961 and 1999, there was a 400% increase in worldwide food exports. Some countries are now economically dependent on food exports, which in some cases account for over 80% of all exports.

In 1994, over 100 countries became signatories to the Uruguay Round of the General Agreement on Tariffs and Trade in a dramatic increase in trade liberalization. This included an agreement to reduce subsidies paid to farmers, underpinned by the WTO enforcement of agricultural subsidy, tariffs, import quotas, and settlement of trade disputes that cannot be bilaterally resolved. Where trade barriers are raised on the disputed grounds of public health and safety, the WTO refer the dispute to the Codex Alimentarius Commission, which was founded in 1962 by the United Nations Food and Agriculture Organization and the World Health Organization. Trade liberalization has greatly affected world food trade.

Marketing and Retailing

Food marketing brings together the producer and the consumer. It is the chain of activities that brings food from "farm gate to plate". The marketing of even a single food product can be a complicated process involving many producers and companies.

For example, fifty-six companies are involved in making one can of chicken noodle soup. These businesses include not only chicken and vegetable processors but also the companies that transport the ingredients and those who print labels and manufacture cans. The food marketing system is the largest direct and indirect non-government employer in the United States.

In the pre-modern era, the sale of surplus food took place once a week when farmers took their wares on market day into the local village marketplace. Here food was sold to grocers for sale in their local shops for purchase by local consumers. With the onset of industrialization and the development of the food processing industry, a wider range of food could be sold and distributed in distant locations. Typically early grocery shops would be counter-based shops, in which purchasers told the shop-keeper what they wanted, so that the shop-keeper could get it for them.

In the 20th century, supermarkets were born. Supermarkets brought with them a self service approach to shopping using shopping carts, and were able to offer quality food at lower cost through economies of scale and reduced staffing costs. In the latter part of the 20th century, this has been further revolutionized by the development of vast warehouse-sized, out-of-town supermarkets, selling a wide range of food from around the world.

Unlike food processors, food retailing is a two-tier market in which a small number of very large companies control a large proportion of supermarkets. The supermarket giants wield great purchasing power over farmers and processors, and strong influence over consumers. Nevertheless, less than 10% of consumer spending on food goes to farmers, with larger percentages going to advertising, transportation, and intermediate corporations.

Prices

It was reported on March 24, 2008, that consumers worldwide faced rising food prices. Reasons for this development include changes in the weather and dramatic changes in the global economy, including higher oil prices, lower food reserves, and growing consumer demand in China and India. In the long term, prices are expected to stabilize. Farmers will grow more grain for both fuel and food and eventually bring prices down. Already this is happening with wheat, with more crops to be planted in the United States, Canada, and Europe in 2009. However, the Food and Agriculture Organization projects that consumers still have to deal with more expensive food until at least 2018.

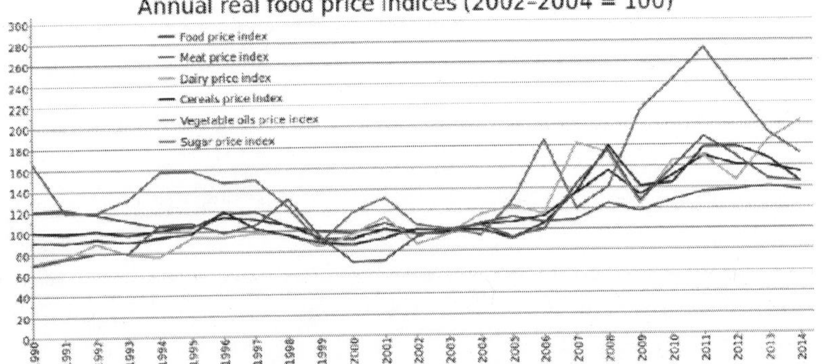

Figure: *Food, meat, dairy, cereals, vegetable oil, and sugar price indices, deflated using the World Bank Manufactures Unit Value Index (MUV).*

It is rare for the spikes to hit all major foods in most countries at once. Food prices rose 4% in the United States in 2007, the highest increase since 1990, and are expected to climb as much again in 2008. As of December 2007, 37 countries faced food crises, and 20 had imposed some sort of food-price controls. In China, the price of pork jumped 58% in 2007. In the 1980s and 1990s, farm subsidies and support programs allowed major grain exporting countries to hold large surpluses, which could be tapped during food shortages to keep prices down. However, new trade policies have made agricultural production much more responsive to market demands, putting global food reserves at their lowest since 1983.

Food prices are rising, wealthier Asian consumers are westernizing their diets, and farmers and nations of the third world are struggling to keep up the pace. The past five years have seen rapid growth in the contribution of Asian nations to the global fluid and powdered milk manufacturing industry, which in 2008 accounted for more than 30%

of production, while China alone accounts for more than 10% of both production and consumption in the global fruit and vegetable processing and preserving industry. The trend is similarly evident in industries such as soft drink and bottled water manufacturing, as well as global cocoa, chocolate, and sugar confectionery manufacturing, forecast to grow by 5.7% and 10.0% respectively during 2008 in response to soaring demand in Chinese and Southeast Asian markets.

Rising food prices over recent years have been linked with social unrest around the world, including rioting in Bangladesh and Mexico, and the Arab Spring.

In 2013 Overseas Development Institute researchers showed that rice has more than doubled in price since 2000, rising by 120% in real terms. This was as a result of shifts in trade policy and restocking by major producers. More fundamental drivers of increased prices are the higher costs of fertiliser, diesel and labour. Parts of Asia see rural wages rise with potential large benefits for the 1.3 billion (2008 estimate) of Asia's poor in reducing the poverty they face. However, this negatively impacts more vulnerable groups who don't share in the economic boom, especially in Asian and African coastal cities. The researchers said the threat means social-protection policies are needed to guard against price shocks. The research proposed that in the longer run, the rises present opportunities to export for Western African farmers with high potential for rice production to replace imports with domestic production.

As Investment

Institutions such as hedge funds, pension funds and investment banks like Barclays Capital, Goldman Sachs and Morgan Stanley have been instrumental in pushing up prices in the last five years, with investment in food commodities rising from $65bn to $126bn (£41bn to £79bn) between 2007 and 2012, contributing to 30-year highs. This has caused price fluctuations which are not strongly related to the actual supply of food, according to the United Nations. Financial institutions now make up 61% of all investment in wheat futures. According to Olivier De Schutter, the UN special rapporteur on food, there was a rush by institutions to enter the food market following George W Bush's Commodities Futures Modernization Act of 2000. De Schutter told the Independent in March 2012: "What we are seeing now is that these financial markets have developed massively with the arrival of these new financial investors, who are purely interested in the short-term monetary gain and are not really interested in the physical thing – they never actually buy the ton of wheat or maize; they only buy a promise to buy or to sell. The result of this financialisation of the commodities market is that the prices of the products respond increasingly to a purely speculative logic. This explains why in very short periods of time we see prices spiking or bubbles exploding, because prices are less and less determined by the real match between supply and demand." In 2011, 450 economists from around the world called on the G20 to regulate the commodities market more.

Some experts have said that speculation has merely aggravated other factors, such as climate change, competition with bio-fuels and overall rising demand. However, some such as Jayati Ghosh, professor of economics at Jawaharlal Nehru University in New Delhi, have pointed out that prices have increased irrespective of supply and demand issues: Ghosh points to world wheat prices, which doubled in the period from June to December 2010, despite there being no fall in global supply.

Famine and Hunger

Food deprivation leads to malnutrition and ultimately starvation. This is often connected with famine, which involves the absence of food in entire communities. This can have a devastating and widespread effect on human health and mortality. Rationing is sometimes used to distribute food in times of shortage, most notably during times of war.

Starvation is a significant international problem. Approximately 815 million people are undernourished, and over 16,000 children die per day from hunger-related causes. Food deprivation is regarded as a deficit need in Maslow's hierarchy of needs and is measured using famine scales.

Exploitation of Natural Resources

The exploitation of natural resources started to emerge in the 19th century as natural resource extraction developed. During the 20th century, energy consumption rapidly increased. Today, about 80% of the world's energy consumption is sustained by the extraction of fossil fuels, which consists of oil, coal and gas. Another non-renewable resource that is exploited by humans are Subsoil minerals such as precious metals that are mainly used in the production of industrial commodities. Intensive agriculture is an example of a mode of production that hinders many aspects of the natural environment, for example the degradation of forests in a terrestrial ecosystem and water pollution in an aquatic ecosystem. As the world population rises and economic growth occurs, the depletion of natural resources influenced by the unsustainable extraction of raw materials becomes an increasing concern.Why Resources are Under Pressure

- Increase in the sophistication of technology enabling natural resources to be extracted quickly and efficiently. E.g., in the past, it could take long hours just to cut down one tree only using saws. Due to increased technology, rates of deforestation have greatly increased
- A rapid increase in population that is now decreasing. The current number of 7.132 billion humans consume many natural resources.
- Cultures of consumerism. Materialistic views lead to the mining of gold and diamonds to produce jewellery, unnecessary commodities for human life or advancement.
- Excessive demand often leads to conflicts due to intense competition. Organizations such as Global Witness and the United Nations have documented the connection.
- Non-equitable distribution of resources.

Problems Arising from the Exploitation of Natural Resources

- Deforestation
- Desertification
- Extinction of species
- Forced migration
- Soil erosion
- Oil depletion
- Ozone depletion
- Greenhouse gas increase
- Extreme energy
- Water pollution
- Natural hazard/Natural disaster

Effects on Local Communities

The Global South

When a mining company enters a developing country to extract raw materials, advocating the advantages of the industry's presence and minimizing the potential negative effects gain cooperation of the local people. Advantageous factors are primarily in economic development so services that the government could not provide such as health centres, police departments and schools can be established. However with economic development, money becomes a dominant subject of interest. This can bring about major conflicts that a local community in a developing country has never dealt with before. These conflicts emerge by a change to more egocentric views among the locals influenced by consumerist values.

The effects of the exploitation of natural resources in the local community of a developing country are exhibited in the impacts from the Ok Tedi Mine. After BHP, now BHP Billiton, entered into Papua New Guinea to exploit copper and gold, the economy of the indigenous peoples boomed. Although their quality of life has improved, initially disputes were common among the locals in terms of land rights and who should be getting the benefits from the mining project. The consequences of the Ok Tedi environmental disaster illustrate the potential negative effects from the exploitation of natural resources. The resulting mining pollution includes toxic contamination of the natural water supply for communities along the Ok Tedi River, causing widespread killing of aquatic life. When a mining company ends a project after extracting the raw materials from an area of a developing country, the local people are left to manage with the environmental damage done to their community and the long run sustainability of the economic benefits stimulated by the mining company's presence becomes a concern.

Resource Curse

The resource curse, also known as the paradox of plenty, refers to the paradox that countries and regions with an abundance of natural resources, specifically point-source non-renewable resources like minerals and fuels, tend to have less economic growth and worse development outcomes than countries with fewer natural resources. This is hypothesized to happen for many different reasons, including a decline in the competitiveness of other economic sectors (caused by appreciation of the real exchange rate as resource revenues enter an economy, a phenomenon known as Dutch disease), volatility of revenues from the natural resource sector due to exposure to global commodity market swings, government mismanagement of resources, or weak, ineffectual, unstable or corrupt institutions (possibly due to the easily diverted actual or anticipated revenue stream from extractive activities). The Resource Curse may not be universal for all countries with an abundance of natural resources, but "for many countries it is real."

Resource Curse Thesis

The idea that natural resources might be more an economic curse than a blessing began to emerge in the 1980s. The term *resource curse thesis* was first used by Richard Auty in 1993 to describe how countries rich in natural resources were unable to use that wealth to boost their economies and how, counter-intuitively, these countries had lower economic growth than countries without an abundance of natural resources. Numerous studies, including one by Jeffrey Sachs and Andrew Warner, have shown a link between natural resource abundance and poor economic growth. This disconnect between natural resource wealth and economic growth can be seen by looking at an example from the petroleum-producing countries. From 1965 to 1998, in the OPEC countries, gross national product per capita growth decreased on average by 1.3%, while in the rest of the developing world, per capita growth was on average 2.2%. Some argue that financial flows from foreign aid can provoke effects that are similar to the resource curse. Abundance of financial resources in absence of sufficient innovation effort in the corporate sector may also lead to the problem of "resource curse."

Negative Effects and Causes

Tug of War Between People and Government

The ambitions of the people and the government conflict, due to the large amount of resources and money a country's government amass for their own luxuries rather than for the people. Thus natural resources serve as a curse for the people, who then have a lower relative standard of living. This may also be caused by the relative greed of the government as they seek to finance their own lives not considering the rest of the people bound to benefit from the resources.

Does an Abundance in Resources Affect the Level of Freedom in a Country?

"Peter Kaznacheev was a senior advisor in the Russian Presidential Administration from 2003 to 2005. He is currently an associate professor and senior research fellow in energy and natural resources at the Russian Academy of National Economy and Public Administration." Kaznacheev is a non-believer in the resource curse. Supporting his claim by stating that the wealthier economically- free countries which are also rich in natural resources don't seem to be "cursed." However, it seems like more than just a coincidence that ""...According to the Index of Economic Freedom, nine out of the 10 least economically free countries are also resource-rich countries."

Internal Conflict

Natural resources can, and often do, provoke conflicts within societies (Collier 2007), as different groups and factions fight for their share. Sometimes these emerge openly as separatist conflicts in regions where the resources are produced (such as in Angola's oil-rich Cabinda province) but often the conflicts occur in more hidden forms, such as fights between different government ministries or departments for access to budgetary allocations. This tends to erode governments' abilities to function effectively. There are several main types of relationships between natural resources and armed conflicts. First, resource curse effects can undermine the quality of governance and economic performances, thereby increasing the vulnerability of countries to conflicts (the 'resource curse' argument). Second, conflicts can occur over the control and exploitation of resources and the allocation of their revenues (the 'resource war' argument). Third, access to resource revenues by belligerents can prolong conflicts (the 'conflict resource' argument). According to one academic study, a country that is otherwise typical but has primary commodity exports around 25% of GDP has a 33% risk of conflict, but when exports are 5% of GDP the chance of conflict drops to 6%.

International war and Conflict

The term "petro-aggression" has been used to describe the tendency of oil-rich states to instigate international conflicts, or to be the targets thereof. There are many examples including: Iraq's invasion of Iran and Kuwait; Libya's repeated incursions into Chad in the 1970s and 1980s; Iran's long-standing suspicion of Western powers; USA's relations with Iraq and Iran.

It is not clear whether the pattern of petro-aggression found in oil-rich countries also applies to other natural resources besides oil.

Taxation

In many economies that are not resource-dependent, governments tax citizens, who demand efficient and responsive government in return. This bargain establishes a political relationship between rulers and subjects. In countries whose economies are dominated by natural resources, however, rulers don't need to tax their citizens because they have a guaranteed source of income from natural resources. Because the country's citizens aren't being taxed, they have less incentive to be watchful with how government spends its money. In addition, those benefiting from mineral resource wealth may perceive an effective and watchful civil service and civil society as a threat to the benefits that they enjoy, and they may take steps to thwart them. As a result, citizens are often poorly served by their rulers, and if the citizens complain, money from the natural resources enables governments to pay for armed forces to keep the citizens in check. Countries whose economies are dominated by resource extraction industries tend to be more repressive, corrupt and badly managed.

Dutch Disease

Dutch disease is an economic phenomenon in which the revenues from natural resource exports damage a nation's productive economic sectors by causing an increase of the real exchange rate and wage increase. This makes tradable sectors, notably agriculture and manufacturing, less competitive in world markets. Absent currency manipulation or a currency peg, appreciation of the currency can damage other sectors, leading to a compensating unfavourable balance of trade. As imports become cheaper, internal employment suffers and with it the skill infrastructure and manufacturing capabilities of the nation. This problem has historically influenced the domestic economics of large empires including Rome during its transition from a Republic, and England during the height of its colonial empire. To compensate for the loss of local employment opportunities, government resources are used to artificially create employment.

The increasing national revenue will often also result in higher government spending on health, welfare, military, and public infrastructure, and if this is done corruptly or inefficiently it can be a burden on the economy. (If it is done efficiently this can boost economic competitiveness - effectively acting as a wage subsidy). While the decrease in the sectors exposed to international competition and consequently even greater dependence on natural resource revenue leaves the economy vulnerable to price changes in the natural resource, this can be managed by an active and effective use of hedge instruments such as forwards, futures, options and swaps, however if it is managed inefficiently or corruptly this can lead to disastrous results. Also, since productivity generally increases faster in the manufacturing sector, the economy will lose out on some of those productivity gains. Dutch Disease first became apparent after the Dutch discovered a massive natural gas field in Groningen in 1959. The Netherlands sought to tap this resource in an attempt to export the gas for profit. However when the gas began to flow out of the country so too did its ability to compete against other countries' exports. With the Netherlands' focus primarily on the new gas exports, the Dutch currency grew at a very quick rate which harmed the country's ability to export other products. With the growing gas market and the shrinking export economy, the Netherlands began to experience a recession. This process has been witnessed in multiple countries around the world including but not limited to Venezuela (oil), Angola (diamonds, oil), the Democratic Republic of the Congo (diamonds), and various other nations. All of these resources are considered "resource-cursed".

The Dutch Disease is Contrasted by the Norwegian Paradox

Revenue Colatility

Prices for some natural resources are subject to wide fluctuation: for example crude oil prices rose from around \$3 per barrel to \$12/bbl in 1974 following the 1973 oil crisis and fell from \$27/bbl to below \$10/bbl during the 1986 glut. In the decade from 1998 to 2008, it rose from \$10/bbl to \$145/bbl, before falling by more than half to \$60/bbl over a few months. When government revenues are dominated by inflows from natural resources (for example, 99.3% of Angola's exports came from just oil and diamonds in 2005), this volatility can play havoc with government planning and debt service. Abrupt changes in economic realities that result from this often provoke widespread breaking of contracts or curtailment of social programs, eroding the rule of law and popular support. Management of this risk is the basis for responsible use of financial hedges.

Excessive Borrowing

Since governments expect more income in the future, they start accumulating debt, even though they are receiving natural resource revenues as well. This is encouraged, since, if the real exchange rate increases, through capital inflows or the Dutch disease, this makes the interest payments on the debt cheaper. In addition, the country's natural resources act as collateral leading to more credit. However, if the natural resources' prices begin to fall, and if the real exchange rate falls, a government would have less money with which to pay a more expensive debt. For example, many oil-rich countries like Nigeria and Venezuela saw rapid expansions of their debt burdens during the 1970s oil boom; however, when oil prices fell in the 1980s, bankers stopped lending to them and many of them fell into arrears, triggering penalty interest charges that made their debts grow even more.

Corruption

In resource-rich countries, it is often easier to maintain authority through allocating resources to favoured constituents than through growth-oriented economic policies and a level, well-regulated playing field. Huge flows of money from natural resources fuel this political corruption. The government has less need to build up the institutional infrastructure to regulate and tax a productive economy outside the resource sector, so the economy may remain undeveloped. The presence of offshore tax havens provide widespread opportunities for corrupt politicians to hide their wealth.

Many extractive operations are illegal and encouraged by corrupt multinational corporations in collusion with national governments. Objections made by indigenous inhabitants are usually ignored. Ed Ayres of Worldwatch Institute reported on these collusive operations in many parts of the world.

The United States Senate Foreign Relations Committee report entitled "Petroleum and Poverty Paradox" states that "too often, oil money that should go to a nation's poor ends up in the pockets of the rich, or it may be squandered on grand palaces and massive showcase projects instead of being invested productively".

In effort to combat resource revenue corruption, a recent report by the Centre for Global Development advocates direct distribution of revenues. Under this proposal, a government would transfer some or all of the revenue from natural resource extraction to citizens in a universal, transparent, and regular payment. Having put this money in the hands of its citizens, the state would treat it like normal income and tax it accordingly—thus forcing the state to collect taxes and fueling public demand for the government to be transparent and accountable in its management of natural resource revenues and in the delivery of public services.

Lack of Diversification and Enclave Effects

Economic diversification may be neglected by authorities or delayed in the light of the temporary high profitability of the limited natural resources. The attempts at diversification that do occur are often grand public works projects which may be misguided or mismanaged. However, even if the authorities try to diversify the economy, this is made difficult because the resource extraction is vastly more lucrative and out-competes other industry. Successful natural-resource-exporting countries often become more dependent on extractive industries over time. The abundance of revenue from natural resources discourages long-term investment in infrastructure which would support a more diverse economy, increasing the negative impact of sudden resource-price drops. While the resource sectors tend to provide large financial revenues, they often provide few jobs, and tend to operate as enclaves with few forward and backward connections to the rest of the economy.

Human Resources

In many poor countries, Natural Resource industries tend to pay far higher salaries than what would be available elsewhere in the economy. This tends to attract the best talent from both private and government sectors, damaging these sectors by depriving them of their best skilled personnel. Another possible effect of the resource curse is the crowding out of human capital; countries that rely on natural resource exports may tend to neglect education because they see no immediate need for it. Resource-poor economies like Singapore, Taiwan or South Korea, by contrast, spent enormous efforts on education, and this contributed in part to their economic success. Other researchers, however, dispute this conclusion; they argue that natural resources generate easily taxable rents that more often than not result in increased spending on education.

Human Rights

It has also been argued that one can correlate rises and falls in the price of petroleum with rises and falls in the implementation of human rights in major oil-producing countries. Human rights throughout resource-cursed countries are dismal or completely lacking. Most normally resource-cursed countries are ruled by either authoritarian or other types of highly repressive regimes. These regimes are kept in power by a select elite such as high-ranking politicians and military leaders. As long as the existing government keeps these few happy they can rule without fear of consequence. This system is set up so that the common folk, those most in need of protection, are left to fend for themselves. In places like the Democratic Republic of the Congo human life has almost no value and slave labour is common. Venezuela demonstrated the resource curse because of its abundance of oil. The country which saw the easy wealth of the oil did not need to expand into other markets due to comparative advantage. The comparative advantage states that if Venezuela can produce oil better than other exports then it should produce the oil. However since Venezuela's oil company is nationally owned it has been criticized as inefficient and backwards. The resource curse is also evident in Venezuela because of the enormous gap between the rich and the poor. Venezuela has a history of oil profits going to wealthy elites with little to none trickling down to the lower socioeconomic classes. The resource curse can cause a country's government to turn volatile towards their people because they do not need them for taxation. The Democratic Republic of the Congo is perhaps one of the most volatile places on the planet, suffering from disease, civil war, and the resource curse. The resource curse of the Congo takes the form of diamonds. While no true government exists each faction exploits the people they rule to harvest diamonds and other rare minerals. The DRC has a unique type of the resource curse because the added factor of its political instability. Instead of fueling a wealthy elite, like many resource-cursed countries, the Congo's diamonds are used to fund the ongoing civil war.

Criticisms

A 2008 study argues that the curse vanishes when looking not at the relative importance of resource exports in the economy but rather at a different measure: the relative abundance of natural resources in the ground. Using that variable to compare countries, it reports that resource wealth in the ground correlates with slightly higher economic growth and slightly fewer armed conflicts. That a high dependency on resource exports correlates with bad policies and effects is not caused by the large degree of resource exportation. The causation goes in the opposite direction: conflicts and bad policies created the heavy dependence on exports of natural resources. When a country's chaos and economic policies scare off foreign investors and send local entrepreneurs abroad to look for better opportunities, the economy becomes skewed. Factories may close and businesses may flee, but petroleum and precious metals remain for the taking. Resource extraction becomes the "default sector" that still functions after other industries have come to a halt.

A 2010 working essay that examines the long-term relationship between natural resource reliance and regime type across the world from 1800 to 2006 demonstrates that increases in natural resource reliance do not induce authoritarianism. On the contrary, the authors find evidence that suggests that increasing reliance on natural resources promotes democratization. These researchers also provide qualitative evidence for this fact across several countries in another article; as well as evidence that there is no relationship between resource reliance and authoritarianism in Latin America.

A 2011 study argues that previous assumptions that oil abundance is a curse were based on methodologies which failed to take into account cross-country differences and dependencies arising from global shocks, such as changes in technology and the price of oil. The researchers studied data from the World Bank over the period 1980 to 2006 for 53 countries, covering 85% of world GDP and 81% of world proven oil reserves. They found that oil abundance positively affected both short-term growth and long-term income levels. In a companion paper, using data on 118 countries over the period 1970-2007, they show that it is the volatility in commodity prices, rather than abundance per se, that drives the resource curse paradox.

Chapter 6

Major Crops

Rice

Rice is the seed of the monocot plants *Oryza sativa* (Asian rice) or *Oryza glaberrima* (African rice). As a cereal grain, it is the most widely consumed staple food for a large part of the world's human population, especially in Asia. It is the agricultural commodity with the third-highest worldwide production, after sugarcane and maize, according to data of FAOSTAT 2012.

Since a large portion of maize crops are grown for purposes other than human consumption, rice is the most important grain with regard to human nutrition and caloric intake, providing more than one fifth of the calories consumed worldwide by humans.

Chinese legends attribute the domestication of rice to Shennong, the legendary Emperor of China and inventor of Chinese agriculture. Genetic evidence has shown that rice originates from a single domestication 8,200–13,500 years ago in the Pearl River valley region of China. Previously, archaeological evidence had suggested that rice was domesticated in the Yangtze River valley region in China. From East Asia, rice was spread to Southeast and South Asia. Rice was introduced to Europe through Western Asia, and to the Americas through European colonization. There are many varieties of rice and culinary preferences tend to vary regionally. In some areas such as the Far East or Spain, there is a preference for softer and stickier varieties.

Rice is normally grown as an annual plant, although in tropical areas it can survive as a perennial and can produce a ratoon crop for up to 30 years. The rice plant can grow to 1–1.8 m (3.3–5.9 ft) tall, occasionally more depending on the variety and soil fertility. It has long, slender leaves 50–100 cm (20–39 in) long and 2–2.5 cm (0.79–0.98 in) broad. The small wind-pollinated flowers are produced in a branched arching to pendulous inflorescence 30–50 cm (12–20 in) long. The edible seed is a grain (caryopsis) 5–12 mm (0.20–0.47 in) long and 2–3 mm (0.079–0.118 in) thick.

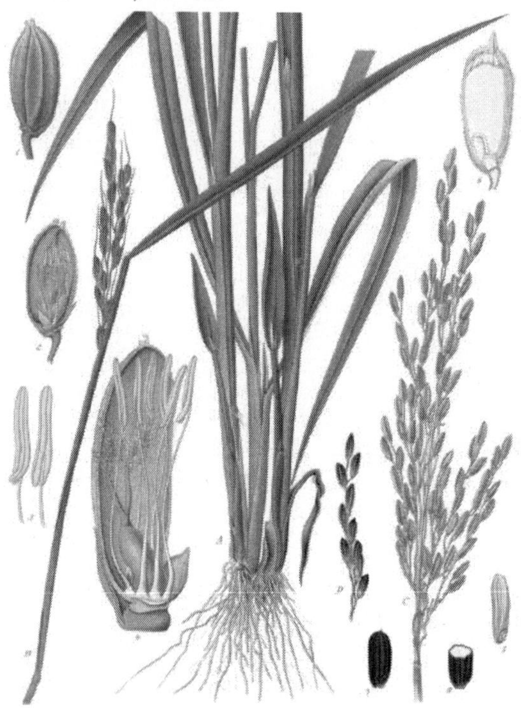

Figure: Oryza sativa, *commonly known as Asian rice*

Rice cultivation is well-suited to countries and regions with low labour costs and high rainfall, as it is labour-intensive to cultivate and requires ample water. However, rice can be grown practically anywhere, even on a steep hill or mountain area with the use of water-controlling terrace systems. Although its parent species are native to Asia and certain parts of Africa, centuries of trade and exportation have made it commonplace in many cultures worldwide.

The traditional method for cultivating rice is flooding the fields while, or after, setting the young seedlings. This simple method requires sound planning and servicing of the water damming and channeling, but reduces the growth of less robust weed and pest plants that have no submerged growth state, and deters vermin. While flooding is not mandatory for the cultivation of rice, all other methods of irrigation require higher effort in weed and pest control during growth periods and a different approach for fertilizing the soil.

The name wild rice is usually used for species of the genera *Zizania* and *Porteresia*, both wild and domesticated, although the term may also be used for primitive or uncultivated varieties of *Oryza*.

Nutrition and Health

Nutrients and the Nutritional Importance of Rice

Rice is the staple food of over half the world's population. It is the predominant dietary energy source for 17 countries in Asia and the Pacific, 9 countries in North and South America and 8 countries in Africa. Rice provides 20% of the world's dietary energy supply, while wheat supplies 19% and maize (corn) 5%.

A detailed analysis of nutrient content of rice suggests that the nutrition value of rice varies based on a number of factors. It depends on the strain of rice, that is between white, brown, black, red and purple varieties of rice – each prevalent in different parts of the world. It also depends on nutrient quality of the soil rice is grown in, whether and how the rice is polished or processed, the manner it is enriched, and how it is prepared before consumption. An illustrative comparison between white and brown rice of protein quality, mineral and vitamin quality, carbohydrate and fat quality suggests that neither is a complete nutrition source. Between the two, there is a significant difference in fibre content and minor differences in other nutrients.

Brilliantly coloured rice strains, such as purple rice, derive their colour from anthocyanins and tocols. Scientific studies suggest that these colour pigments have antioxidant properties that may be useful to human health. In purple rice bran, hydrophilic antioxidants are in greater quantity and have higher free radical scavenging activity than lipophilic antioxidants. Anthocyanins and γ-tocols in purple rice are largely located in the inner portion of purple rice bran.

Comparative nutrition studies on red, black and white varieties of rice suggest that pigments in red and black rice varieties may offer nutritional benefits. Red or black rice consumption was found to reduce or retard the progression of atherosclerotic plaque development, induced by dietary cholesterol, in mammals. White rice consumption offered no similar benefits, and the study claims this to be due to absent antioxidants in red and black varieties of rice.

Rice-growing Environments

Rice can be grown in different environments, depending upon water availability. Generally, rice does not thrive in a waterlogged area, yet it can survive and grow herein and it can also survive flooding.

1. Lowland, rainfed, which is drought prone, favours medium depth; waterlogged, submergence, and flood prone
2. Lowland, irrigated, grown in both the wet season and the dry season
3. Deep water or floating rice
4. Coastal Wetland
5. Upland rice is also known as Ghaiya rice, well known for its drought tolerance

History of Domestication and Cultivation

There have been plenty of debates on the origins of the domesticated rice. Genetic evidence published in the *Proceedings of the National Academy of Sciences of the United States of America* (PNAS) shows that all forms of Asian rice, both *indica* and *japonica*, spring from a single domestication that occurred 8,200–13,500 years ago in China of the wild rice *Oryza rufipogon*. A 2012 study published in *Nature*, through a map of rice genome variation, indicated that the domestication of rice occurred in the Pearl River valley region of China based on the genetic evidence. From East Asia, rice was spread to South and Southeast Asia. Before this research, the commonly accepted view, based on archaeological evidence, is that rice was first domesticated in the region of the Yangtze River valley in China.

Morphological studies of rice phytoliths from the Diaotonghuan archaeological site clearly show the transition from the collection of wild rice to the cultivation of domesticated rice. The large number of wild rice phytoliths at the Diaotonghuan level dating from 12,000–11,000 BP indicates that wild rice collection was part of the local means of subsistence. Changes in the morphology of Diaotonghuan phytoliths dating from 10,000–8,000 BP show that rice had by this time been domesticated. Soon afterwards the two major varieties of indica and japonica rice were being grown in Central China. In the late 3rd millennium BC, there was a rapid expansion of rice cultivation into mainland Southeast Asia and westwards across India and Nepal.

In 2003, Korean archaeologists claimed to have discovered the world's oldest domesticated rice. Their 15,000-year old age challenges the accepted view that rice cultivation originated in China about 12,000 years ago. These findings were received by academia with strong skepticism, and the results and their publicizing has been cited as being driven by a combination of nationalist and regional interests. In 2011, a combined effort by the Stanford University, New York University, Washington University in St. Louis, and Purdue University has provided the strongest evidence yet that there is only one single origin of domesticated rice, in the Yangtze Valley of China.

The earliest remains of the grain in the Indian subcontinent have been found in the Indo-Gangetic Plain and date from 7000–6000 BC though the earliest widely accepted date for cultivated rice is placed at around 3000–2500 BC with findings in regions belonging to the Indus Valley Civilization. Perennial wild rices still grow in Assam and Nepal. It seems to have appeared around 1400 BC in southern India after its domestication in the northern plains. It then spread to all the fertile alluvial plains watered by rivers. Cultivation and cooking methods are thought to have spread to the west rapidly and by medieval times, southern Europe saw the introduction of rice as a hearty grain.

Rice spread to the Middle East where, according to Zohary and Hopf (2000, p. 91), *O. sativa* was recovered from a grave at Susa in Iran (dated to the 1st century AD).

Regional History

In a recent study, scientist have found a link for differences in human culture based on either wheat or rice cultivating races since ancient times.

Africa

African rice has been cultivated for 3500 years. Between 1500 and 800 BC, *Oryza glaberrima* propagated from its original centre, the Niger River delta, and extended to Senegal. However, it never developed far from its original region. Its cultivation even declined in favour of the Asian species, which was introduced to East Africa early in the common era and spread westward. African rice helped Africa conquer its famine of 1203.

Asia

Today, the majority of all rice produced comes from China, India, Indonesia, Bangladesh, Vietnam, Thailand, Myanmar, Pakistan, Philippines, Korea and Japan. Asian farmers still account for 87% of the world's total rice production.

Sri Lanka

Rice is the staple food amongst all the ethnic groups in Sri Lanka. Agriculture in Sri Lanka mainly depends on the rice cultivation. Rice production is acutely dependent on rainfall and government supply necessity of water through irrigation channels throughout the cultivation seasons. The principal cultivation season, known as "Maha", is from October to March and the subsidiary cultivation season, known as "Yala", is from April to September. During Maha season, there is usually enough water to sustain the cultivation of all rice fields, nevertheless in Yala season there is only enough water for cultivation of half of the land extent.

Traditional rice varieties are now making a comeback with the recent interest in green foods.

Thailand

Rice is the main export of Thailand, especially the white jasmine rice 105 (Dok Mali 105). Thailand has a large number of rice varieties, 3,500 kinds with different characters, and 5 kinds of wild rice cultivates. In each region of the country there are different rice seed types. Their use depends on weather, atmosphere, and topography.

The northern region has both low lands and high lands. The farmers' usual crop is non-glutinous rice such as Niew Sun Pah Tong rice seeds. This rice is naturally protected from leaf disease, and the paddy has a brown colour. The northeastern region has a large area, where farmers can cultivate about 36 million square metres of rice. Although most of them are plains and dry areas, they can grow the white jasmine rice 105 which is the most famous Thai rice. The white jasmine rice was developed in Chonburi province first and after that it was grown in many areas in the country but the rice from this region has a high quality, because it's softer, whiter and more fragrant. This rice can resist drought, acidic soil, and alkaline soil.

The central region is mostly composed of plains. Most farmers grow Jao rice. For example the Pathum Thani 1 rice which has qualities similar to the white jasmine 105 rice. Their paddy has the colour of thatch and their cooked rice has fragrant grains also.

In the southern region, most farmers transplant around boundaries to the flood of plain or plain between mountains. Farming is the region is slower than other regions because the rainy season comes late. The popular rice varieties in this area are the Leb Nok Pattani seeds, a type of Jao rice. Their paddy has the colour of thatch and it can be processed to make noodles.

Companion Plant

One of the earliest known examples of companion planting is the growing of rice with Azolla, the mosquito fern, which covers the top of a fresh rice paddy's water, blocking out any competing plants, as well as fixing nitrogen from the atmosphere for the rice to use. The rice is planted when it is tall enough to poke out above the azolla. This method has been used for at least a thousand years.

Middle East

Rice was grown in some areas of southern Iraq. With the rise of Islam it moved north to Nisibin, the southern shores of the Caspian Sea(Iran) and then beyond the Muslim world into the valley of the Volga. In Egypt, rice is mainly grown in the Nile Delta. In Israel, rice came to be grown in the Jordan Valley. Rice is also grown in Saudi Arabia at Al-hasa Oasis and in Yemen.

Europe

Rice was known to the Classical world, being imported from Egypt, and perhaps west Asia. It was known to Greece by returning soldiers from Alexander the Great's military expedition to Asia. Large deposits of rice from the first century A.D. have been found in Roman camps in Germany.

The Moors brought Asiatic rice to the Iberian Peninsula in the 10th century. Records indicate it was grown in Valencia and Majorca. In Majorca, rice cultivation seems to have stopped after the Christian conquest, although historians are not certain.

Muslims also brought rice to Sicily, where it was an important crop long before it is noted in the plain of Pisa (1468) or in the Lombard plain (1475), where its cultivation was promoted by Ludovico Sforza, Duke of Milan, and demonstrated in his model farms.

After the 15th century, rice spread throughout Italy and then France, later propagating to all the continents during the age of European exploration.

In European Russia, a short-grain, starchy rice similar to the Italian varieties, has been grown in the Krasnodar Krai, and known in Russia as "Kuban Rice" or "Krasnodar Rice". In the Russian Far East several *japonica* cultivars are grown in Primorye around the Khanka lake. Increasing scale of rice production in the region has recently brought criticism towards growers' alleged bad practices in regards to the environment.

Caribbean and Latin America

Rice is not native to the Americas but was introduced to Latin America and the Caribbean by European colonizers at an early date with Spanish colonizers introducing Asian rice to Mexico in the 1520s at Veracruz and the Portuguese and their African slaves introducing it at about the same time to colonial Brazil. Recent scholarship suggests that enslaved Africans played an active role in the establishment of rice in the New World and that African rice was an important crop from an early period. Varieties of rice and bean dishes that were a staple dish along the peoples of West Africa remained a staple among their descendants subjected to slavery in the Spanish New World colonies, Brazil and elsewhere in the Americas.

The Native Americans of what is now the Eastern United States may have practiced extensive agriculture with forms of wild rice (Zizania palustris), which looks similar to but it not directly related to rice.

United States

In 1694, rice arrived in South Carolina, probably originating from Madagascar.

In the United States, colonial South Carolina and Georgia grew and amassed great wealth from the slave labour obtained from the Senegambia area of West Africa and from coastal Sierra Leone. At the port of Charleston, through which 40% of all American slave imports passed, slaves from this region of Africa brought the highest prices due to their prior knowledge of rice culture, which was put to use on the many rice plantations around Georgetown, Charleston, and Savannah.

From the enslaved Africans, plantation owners learned how to dyke the marshes and periodically flood the fields. At first the rice was laboriously milled by hand using large mortars and pestles made of wood, then winnowed in sweetgrass baskets (the making of which was another skill brought by slaves from Africa). The invention of the rice mill increased profitability of the crop, and the addition of water power for the mills in 1787 by millwright Jonathan Lucas was another step forward.

Rice culture in the southeastern U.S. became less profitable with the loss of slave labour after the American Civil War, and it finally died out just after the turn of the 20th century. Today, people can visit the only remaining rice plantation in South Carolina that still has the original winnowing barn and rice mill from the mid-19th century at the historic Mansfield Plantation in Georgetown, South Carolina. The predominant strain of rice in the Carolinas was from Africa and was known as "Carolina Gold." The cultivar has been preserved and there are current attempts to reintroduce it as a commercially grown crop.

In the southern United States, rice has been grown in southern Arkansas, Louisiana, and east Texas since the mid-19th century. Many Cajun farmers grew rice in wet marshes and low lying prairies where they could also farm crayfish when the fields were flooded. In recent years rice production has risen in North America, especially in the Mississippi embayment in the states of Arkansas and Mississippi.

Rice cultivation began in California during the California Gold Rush, when an estimated 40,000 Chinese labourers immigrated to the state and grew small amounts of the grain for their own consumption. However, commercial production began only in 1912 in the town of Richvale in Butte County. By 2006, California produced the second largest rice crop in the United States, after Arkansas, with production concentrated in six counties north of Sacramento. Unlike the Arkansas–Mississippi Delta region, California's production is dominated by short- and medium-grain *japonica* varieties, including cultivars developed for the local climate such as Calrose, which makes up as much as 85% of the state's crop.

References to wild rice in the Americas are to the unrelated *Zizania palustris*

More than 100 varieties of rice are commercially produced primarily in six states (Arkansas, Texas, Louisiana, Mississippi, Missouri, and California) in the U.S. According to estimates for the 2006 crop year, rice production in the U.S. is valued at $1.88 billion, approximately half of which is expected to be exported. The U.S. provides about 12% of world rice trade. The majority of domestic utilization of U.S. rice is direct food use (58%), while 16% is used in each of processed foods and beer. 10% is found in pet food.

Australia

Rice was one of the earliest crops planted in Australia by British settlers, who had experience with rice plantations in the Americas and India. Although attempts to grow rice in the well-watered north of Australia have been made for many years, they have consistently failed because of inherent iron and manganese toxicities in the soils and destruction by pests.

In the 1920s it was seen as a possible irrigation crop on soils within the Murray-Darling Basin that were too heavy for the cultivation of fruit and too infertile for wheat.

Because irrigation water, despite the extremely low runoff of temperate Australia, was (and remains) very cheap, the growing of rice was taken up by agricultural groups over the following decades. Californian varieties of rice were found suitable for the climate in the Riverina, and the first mill opened at Leeton in 1951.

Monthly Value (A$ Millions) of Rice Imports to Australia Since 1988

Even before this Australia's rice production greatly exceeded local needs, and rice exports to Japan have become a major source of foreign currency. Above-average rainfall from the 1950s to the middle 1990s encouraged the expansion of the Riverina rice industry, but its prodigious water use in a practically waterless region began to attract the attention of environmental scientists. These became severely concerned with declining flow in the Snowy River and the lower Murray River.

Although rice growing in Australia is highly profitable due to the cheapness of land, several recent years of severe drought have led many to call for its elimination because of its effects on extremely fragile aquatic ecosystems. The Australian rice industry is somewhat opportunistic, with the area planted varying significantly from season to season depending on water allocations in the Murray and Murrumbidgee irrigation regions.

Production and Commerce

Production

The world dedicated 162.3 million hectares in 2012 for rice cultivation and the total production was about 738.1 million tonnes. The average world farm yield for rice was 4.5 tonnes per hectare, in 2012. Rice farms in Egypt were the most productive in 2012, with a nationwide average of 9.5 tonnes per hectare. Second place: Australia - 8.9 tonnes per hectare. Third place: USA - 8.3 tonnes per hectare.

Rice is a major food staple and a mainstay for the rural population and their food security. It is mainly cultivated by small farmers in holdings of less than 1 hectare. Rice is also a wage commodity for workers in the cash crop or non-agricultural sectors. Rice is vital for the nutrition of much of the population in Asia, as well as in Latin America and the Caribbean and in Africa; it is central to the food security of over half the world population. Developing countries account for 95% of the total production, with China and India alone responsible for nearly half of the world output.

World production of rice has risen steadily from about 200 million tonnes of paddy rice in 1960 to over 678 million tonnes in 2009. The three largest producers of rice in 2009 were China (197 million tonnes), India (131 Mt), and Indonesia (64 Mt). Among the six largest rice producers, the most productive farms for rice, in 2009, were in China producing 6.59 tonnes per hectare. At 44 million hectares, India had the largest farm area under rice production in 2009. The rice farm productivity in India was about 45% of the rice farm productivity in China, and about 60% of the rice farm productivity in Indonesia.

If India could adopt the farming knowledge and technology in use in China and Indonesia, India could produce an additional 100 million tonnes of rice, enough staple food for about 400 million people every year, and US$50 billion in additional annual income to its rice farmers (adjusted to 2010 dollars and global rice prices per tonne). In the 1990s, genetic studies took place in many European laboratories to increase rice production per hectare. Most of them were Dutch agricultural organizations united by HNGAC. These studies were later stopped due to lack of funding.

In addition to the gap in farming system technology and knowledge, many rice grain producing countries have significant losses post-harvest at the farm and because of poor roads, inadequate storage technologies, inefficient supply chains and farmer's inability to bring the produce into retail markets dominated by small shopkeepers. A World Bank – FAO study claims 8% to 26% of rice is lost in developing nations, on average, every year, because of post-harvest problems and poor infrastructure. Some sources claim the post-harvest losses to exceed 40%. Not only do these losses reduce food security in the world, the study claims that farmers in developing countries such as China, India and others lose approximately US$89 billion of income in preventable post-harvest farm losses, poor transport, the lack of proper storage and retail. One study claims that if these post-harvest grain losses could be eliminated with better infrastructure and retail network, in India alone enough food would be saved every year to feed 70 to 100 million people over a year. However, other writers have warned against dramatic assessments of post-harvest food losses, arguing that "worst-case scenarios" tend to be used rather than realistic averages and that in many cases the cost of avoiding losses exceeds the value of the food saved.

The seeds of the rice plant are first milled using a rice huller to remove the chaff (the outer husks of the grain). At this point in the process, the product is called brown rice. The milling may be continued, removing the bran, *i.e.*, the rest of the husk and the germ, thereby creating white rice. White rice, which keeps longer, lacks some important nutrients; moreover, in a limited diet which does not supplement the rice, brown rice helps to prevent the disease beriberi.

Either by hand or in a rice polisher, white rice may be buffed with glucose or talc powder (often called polished rice, though this term may also refer to white rice in general), parboiled, or processed into flour. White rice may also be enriched by adding nutrients, especially those lost during the milling process. While the cheapest method of enriching involves adding a powdered blend of nutrients that will easily wash off (in the United States, rice which has been so treated requires a label warning against rinsing), more sophisticated methods apply nutrients directly to the grain, coating the grain with a water-insoluble substance which is resistant to washing.

In some countries, a popular form, parboiled rice, is subjected to a steaming or parboiling process while still a brown rice grain. This causes nutrients from the outer husk, especially thiamine, to move into the grain itself. The parboil process causes a gelatinisation of the starch in the grains. The grains become less brittle, and the colour of the milled grain changes from white to yellow.

The rice is then dried, and can then be milled as usual or used as brown rice. Milled parboiled rice is nutritionally superior to standard milled rice. Parboiled rice has an additional benefit in that it does not stick to the pan during cooking, as happens when cooking regular white rice. This type of rice is eaten in parts of India and countries of West Africa are also accustomed to consuming parboiled rice.

Despite the hypothetical health risks of talc (such as stomach cancer), talc-coated rice remains the norm in some countries due to its attractive shiny appearance, but it has been banned in some, and is no longer widely used in others (such as the United States). Even where talc is not used, glucose, starch, or other coatings may be used to improve the appearance of the grains.

Rice bran, called *nuka* in Japan, is a valuable commodity in Asia and is used for many daily needs. It is a moist, oily inner layer which is heated to produce oil. It is also used as a pickling bed in making rice bran pickles and *takuan*.

Raw rice may be ground into flour for many uses, including making many kinds of beverages, such as *amazake, horchata*, rice milk, and rice wine. Rice flour does not contain gluten, so is suitable for people on a gluten-free diet. Rice may also be made into various types of noodles. Raw, wild, or brown rice may also be consumed by raw-foodist or fruitarians if soaked and sprouted (usually a week to 30 days – gaba rice).

Processed rice seeds must be boiled or steamed before eating. Boiled rice may be further fried in cooking oil or butter (known as fried rice), or beaten in a tub to make *mochi*.

Rice is a good source of protein and a staple food in many parts of the world, but it is not a complete protein: it does not contain all of the essential amino acids in sufficient amounts for good health, and should be combined with other sources of protein, such as nuts, seeds, beans, fish, or meat.

Rice, like other cereal grains, can be puffed (or popped). This process takes advantage of the grains' water content and typically involves heating grains in a special chamber. Further puffing is sometimes accomplished by processing puffed pellets in a low-pressure chamber. The ideal gas law means either lowering the local pressure or raising the water temperature results in an increase in volume prior to water evaporation, resulting in a puffy texture. Bulk raw rice density is about 0.9 g/cm^3. It decreases to less than one-tenth that when puffed.

Harvesting, Drying and Milling

Unmilled rice, known as paddy (Indonesia and Malaysia: padi; Philippines, palay), is usually harvested when the grains have a moisture content of around 25%.

In most Asian countries, where rice is almost entirely the product of smallholder agriculture, harvesting is carried out manually, although there is a growing interest in mechanical harvesting. Harvesting can be carried out by the farmers themselves, but is also frequently done by seasonal labour groups. Harvesting is followed by threshing, either immediately or within a day or two. Again, much threshing is still carried out by hand but there is an increasing use of mechanical threshers. Subsequently, paddy needs to be dried to bring down the moisture content to no more than 20% for milling.

A familiar sight in several Asian countries is paddy laid out to dry along roads. However, in most countries the bulk of drying of marketed paddy takes place in mills, with village-level drying being used for paddy to be consumed by farm families. Mills either sun dry or use mechanical driers or both. Drying has to be carried out quickly to avoid the formation of moulds.

Mills range from simple hullers, with a throughput of a couple of tonnes a day, that simply remove the outer husk, to enormous operations that can process 4,000 tonnes a day and produce highly polished rice. A good mill can achieve a paddy-to-rice conversion rate of up to 72% but smaller, inefficient mills often struggle to achieve 60%. These smaller mills often do not buy paddy and sell rice but only service farmers who want to mill their paddy for their own consumption.

Distribution

Because of the importance of rice to human nutrition and food security in Asia, the domestic rice markets tend to be subject to considerable state involvement. While the private sector plays a leading role in most countries, agencies such as BULOG in Indonesia, the NFA in the Philippines, VINAFOOD in Vietnam and the Food Corporation of India are all heavily involved in purchasing of paddy from farmers or rice from mills and in distributing rice to poorer people. BULOG and NFA monopolise rice imports into their countries while VINAFOOD controls all exports from Vietnam.

Trade

World trade figures are very different from those for production, as less than 8% of rice produced is traded internationally. In economic terms, the global rice trade was a small fraction of 1% of world mercantile trade. Many countries consider rice as a strategic food staple, and various governments subject its trade to a wide range of controls and interventions.

Developing countries are the main players in the world rice trade, accounting for 83% of exports and 85% of imports. While there are numerous importers of rice, the exporters of rice are limited. Just five countries – Thailand, Vietnam, China, the United States and India – in decreasing order of exported quantities, accounted for about three-quarters of world rice exports in 2002. However, this ranking has been rapidly changing in recent years. In 2010, the three largest exporters of rice, in decreasing order of quantity exported were Thailand, Vietnam and India. By 2012, India became the largest exporter of rice with a 100% increase in its exports on year to year basis, and Thailand slipped to third position. Together, Thailand, Vietnam and India accounted for nearly 70% of the world rice exports.

The primary variety exported by Thailand and Vietnam were Jasmine rice, while exports from India included aromatic Basmati variety. China, an exporter of rice in early 2000s, was a net importer of rice in 2010 and will become the largest net importer, surpassing Nigeria, in 2013. According to a USDA report, the world's largest exporters of rice in 2012 were India (9.75 million tonnes), Vietnam (7 million tonnes), Thailand (6.5 million tonnes), Pakistan (3.75 million tonnes) and the United States (3.5 million tonnes).

Major importers usually include Nigeria, Indonesia, Bangladesh, Saudi Arabia, Iran, Iraq, Malaysia, the Philippines, Brazil and some African and Persian Gulf countries. In common with other West African countries, Nigeria is actively promoting domestic production. However, its very heavy import duties (110%) open it to smuggling from neighbouring countries. Parboiled rice is particularly popular in Nigeria. Although China and India are the two largest producers of rice in the world, both countries consume the majority of the rice produced domestically, leaving little to be traded internationally.

World's Most Productive Rice Farms and Farmers

The average world yield for rice was 4.3 tonnes per hectare, in 2010. Australian rice farms were the most productive in 2010, with a nationwide average of 10.8 tonnes per hectare. Yuan Longping of China National Hybrid Rice Research and Development Centre, China, set a world record for rice yield in 2010 at 19 tonnes per hectare on a demonstration plot.

In 2011, this record was surpassed by an Indian farmer, Sumant Kumar, with 22.4 tonnes per hectare in Bihar. Both these farmers claim to have employed newly developed rice breeds and System of Rice Intensification (SRI), a recent innovation in rice farming. SRI is claimed to have set new national records in rice yields, within the last 10 years, in many countries. The claimed Chinese and Indian yields have yet to be demonstrated on seven-hectare lots and to be reproducible over two consecutive years on the same farm.

Price

In late 2007 to May 2008, the price of grains rose greatly due to droughts in major producing countries (particularly Australia), increased use of grains for animal feed and US subsidies for bio-fuel production. Although there was no shortage of rice on world markets this general upward trend in grain prices led to panic buying by consumers, government rice export bans (in particular, by Vietnam and India) and inflated import orders by the Philippines marketing board, the National Food Authority. This caused significant rises in rice prices. In late April 2008, prices hit 24 US cents a pound, twice the price of seven months earlier. Over the period of 2007 to 2013, the Chinese government has substantially increased the price it pays domestic farmers for their rice, rising to US$500 per metric ton by 2013. The 2013 price of rice originating from other southeast Asian countries was a comparably low US$350 per metric ton.

On April 30, 2008, Thailand announced plans for the creation of the Organisation of Rice Exporting Countries (OREC) with the intention that this should develop into a price-fixing cartel for rice. However, little progress had been made by mid-2011 to achieve this.

Worldwide Consumption

As of 2009 world food consumption of rice was 531.6 million metric tons of paddy equivalent (354,603 of milled equivalent), while the far largest consumers were China consuming 156.3 million metric tons of paddy equivalent (29.4% of the world consumption) and India consuming 123.5 million metric tons of paddy equivalent (23.3% of the world consumption). Between 1961 and 2002, per capita consumption of rice increased by 40%.

Rice is the most important crop in Asia. In Cambodia, for example, 90% of the total agricultural area is used for rice production.

U.S. rice consumption has risen sharply over the past 25 years, fuelled in part by commercial applications such as beer production. Almost one in five adult Americans now report eating at least half a serving of white or brown rice per day.

Environmental Impacts

Rice cultivation on wetland rice fields is thought to be responsible for 1.5% of the anthropogenic methane emissions. Rice requires slightly more water to produce than other grains. Rice production uses almost a third of Earth's fresh water.

Long-term flooding of rice fields cuts the soil off from atmospheric oxygen and causes anaerobic fermentation of organic matter in the soil. Methane production from rice cultivation contributes ~1.5% of anthropogenic greenhouse gases. Methane is twenty times more potent a greenhouse gas than carbon dioxide.

A 2010 study found that, as a result of rising temperatures and decreasing solar radiation during the later years of the 20th century, the rice yield growth rate has decreased in many parts of Asia, compared to what would have been observed had the temperature and solar radiation trends not occurred. The yield growth rate had fallen 10–20% at some locations. The study was based on records from 227 farms in Thailand, Vietnam, Nepal, India, China, Bangladesh, and Pakistan. The mechanism of this falling yield was not clear, but might involve increased respiration during warm nights, which expends energy without being able to photosynthesize.

Wheat

Wheat (*Triticum* spp.) is a cereal grain, originally from the Levant region of the Near East but now cultivated worldwide. In 2010, world production of wheat was 651 million tons, making it the third most-produced cereal after maize (844 million tons) and rice (672 million tons). Wheat was the second most-produced cereal in 2009; world production in that year was 682 million tons, after maize (817 million tons), and with rice as a close third (679 million tons).

This grain is grown on more land area than any other commercial food. World trade in wheat is greater than for all other crops combined. Globally, wheat is the leading source of vegetable protein in human food, having a higher protein content than other major cereals, maize (corn) or rice. In terms of total production tonnages used for food, it is currently second to rice as the main human food crop and ahead of maize, after allowing for maize's more extensive use in animal feeds.

Wheat was a key factor enabling the emergence of city-based societies at the start of civilization because it was one of the first crops that could be easily cultivated on a large scale, and had the additional advantage of yielding a harvest that provides long-term storage of food. Wheat contributed to the emergence of city-states in the Fertile Crescent, including the Babylonian and Assyrian empires. Wheat grain is a staple food used to make flour for leavened, flat and steamed breads, biscuits, cookies, cakes, breakfast cereal, pasta, noodles, couscous and for fermentation to make beer, other alcoholic beverages, or biofuel.

There are six wheat classifications: 1) hard red winter, 2) hard red spring, 3) soft red winter, 4) durum (hard), 5) Hard white, 6) soft white wheat. The hard wheats have the most amount of gluten and are used for making bread, rolls and all-purpose flour. The soft wheats are used for making flat bread, cakes, pastries, crackers, muffins, and biscuits.

Wheat is planted to a limited extent as a forage crop for livestock, although the straw cannot be used as feed. Its straw can be used as a construction material for roofing thatch. The whole grain can be milled to leave just the endosperm for white flour. The by-products of this are bran and germ. The whole grain is a concentrated source of vitamins, minerals, and protein, while the refined grain is mostly starch.

Major Cultivated Species of Wheat

Hexaploid Species

- Common wheat or Bread wheat (*T. aestivum*) – A hexaploid species that is the most widely cultivated in the world.
- Spelt (*T. spelta*) – Another hexaploid species cultivated in limited quantities. Spelt is sometimes considered a subspecies of the closely related species common wheat (*T. aestivum*), in which case its botanical name is considered to be *Triticum aestivum* subsp. *spelta*.

Tetraploid Species

- Durum (*T. durum*) – The only tetraploid form of wheat widely used today, and the second most widely cultivated wheat.
- Emmer (*T. dicoccon*) – A tetraploid species, cultivated in ancient times but no longer in widespread use.

Diploid Species

- Einkorn (*T. monococcum*) – A diploid species with wild and cultivated variants. Domesticated at the same time as emmer wheat, but never reached the same importance.

Classes used in the United States:

- Durum – Very hard, translucent, light-coloured grain used to make semolina flour for pasta & bulghur; high in protein, specifically, gluten protein.
- Hard Red Spring – Hard, brownish, high-protein wheat used for bread and hard baked goods. Bread Flour and high-gluten flours are commonly made from hard red spring wheat. It is primarily traded at the Minneapolis Grain Exchange.
- Hard Red Winter – Hard, brownish, mellow high-protein wheat used for bread, hard baked goods and as an adjunct in other flours to increase protein in pastry flour for pie crusts. Some brands of unbleached all-purpose flours are commonly made from hard red winter wheat alone. It is primarily traded on the Kansas City Board of Trade. One variety is known as "turkey red wheat", and was brought to Kansas by Mennonite immigrants from Russia.
- Soft Red Winter – Soft, low-protein wheat used for cakes, pie crusts, biscuits, and muffins. Cake flour, pastry flour, and some self-rising flours with baking powder and salt added, for example, are made from soft red winter wheat. It is primarily traded on the Chicago Board of Trade.
- Hard White – Hard, light-coloured, opaque, chalky, medium-protein wheat planted in dry, temperate areas. Used for bread and brewing.
- Soft White – Soft, light-coloured, very low protein wheat grown in temperate moist areas. Used for pie crusts and pastry. Pastry flour, for example, is sometimes made from soft white winter wheat.

Red wheats may need bleaching; therefore, white wheats usually command higher prices than red wheats on the commodities market.

As a Food

Raw wheat can be ground into flour or, using hard durum wheat only, can be ground into semolina; germinated and dried creating malt; crushed or cut into cracked wheat; parboiled (or steamed), dried, crushed and de-branned into bulgur also known as groats. If the raw wheat is broken into parts at the mill, as is usually done, the outer husk or bran can be used several ways. Wheat is a major ingredient in such foods as bread, porridge, crackers, biscuits, Muesli, pancakes, pies, pastries, cakes, cookies, muffins, rolls, doughnuts, gravy, boza (a fermented beverage), and breakfast cereals (e.g., Wheatena, Cream of Wheat, Shredded Wheat, and Wheaties).

Nutrition

100 g (3.5 oz) of hard red winter wheat contain about 12.6 g (0.44 oz) of protein, 1.5 g (0.053 oz) of total fat, 71 g (2.5 oz) of carbohydrate (by difference), 12.2 g (0.43 oz) of dietary fibre, and 3.2 mg (0.00011 oz) of iron (17% of the daily requirement); the same weight of hard red spring wheat contains about 15.4 g (0.54 oz) of protein, 1.9 g (0.067 oz) of total fat, 68 g (2.4 oz) of carbohydrate (by difference), 12.2 g (0.43 oz) of dietary fibre, and 3.6 mg (0.00013 oz) of iron (20% of the daily requirement).

Much of the carbohydrate fraction of wheat is starch. Wheat starch is an important commercial product of wheat, but second in economic value to wheat gluten. The principal parts of wheat flour are gluten and starch. These can be separated in a kind of home experiment, by mixing flour and water to form a small ball of dough, and kneading it gently while rinsing it in a bowl of water. The starch falls out of the dough and sinks to the bottom of the bowl, leaving behind a ball of gluten.

In wheat, phenolic compounds are mainly found in the form of insoluble bound ferulic acid and are relevant to resistance to wheat fungal diseases. Alkylresorcinols are phenolic lipids present in high amounts in the bran layer (e.g. pericarp, testa and aleurone layers) of wheat and rye (0.1-0.3% of dry weight).

Worldwide Consumption

Wheat is grown on more than 216,000,000 hectares (530,000,000 acres), larger than for any other crop. World trade in wheat is greater than for all other crops combined. With rice, wheat is the world's most favoured staple food. It is a major diet component because of the wheat plant's agronomic adaptability with the ability to grow from near arctic regions to equator, from sea level to plains of Tibet, approximately

4,000 m (13,000 ft) above sea level. In addition to agronomic adaptability, wheat offers ease of grain storage and ease of converting grain into flour for making edible, palatable, interesting and satisfying foods. Wheat is the most important source of carbohydrate in a majority of countries.

Wheat protein is easily digested by nearly 99% of human population, as is its starch. Wheat also contains a diversity of minerals, vitamins and fats (lipids). With a small amount of animal or legume protein added, a wheat-based meal is highly nutritious.

The most common forms of wheat are white and red wheat. However, other natural forms of wheat exist. For example, in the highlands of Ethiopia grows purple wheat, a tetraploid species of wheat that is rich in anti-oxidants. Other commercially minor but nutritionally promising species of naturally evolved wheat species include black, yellow and blue wheat.

Commercial Use

Harvested wheat grain that enters trade is classified according to grain properties for the purposes of the commodity markets. Wheat buyers use these to decide which wheat to buy, as each class has special uses, and producers use them to decide which classes of wheat will be most profitable to cultivate.

Wheat is widely cultivated as a cash crop because it produces a good yield per unit area, grows well in a temperate climate even with a moderately short growing season, and yields a versatile, high-quality flour that is widely used in baking. Most breads are made with wheat flour, including many breads named for the other grains they contain like most rye and oat breads. The popularity of foods made from wheat flour creates a large demand for the grain, even in economies with significant food surpluses.

In recent years, low international wheat prices have often encouraged farmers in the USA to change to more profitable crops. In 1998, the price at harvest was $2.68 per bushel. USDA report revealed that in 1998, average operating costs were $1.43 per bushel and total costs were $3.97 per bushel. In that study, farm wheat yields averaged 41.7 bushels per acre (2.2435 metric ton/hectare), and typical total wheat production value was $31,900 per farm, with total farm production value (including other crops) of $173,681 per farm, plus $17,402 in government payments. There were significant profitability differences between low- and high-cost farms, mainly due to crop yield differences, location, and farm size.

In 2007 there was a dramatic rise in the price of wheat due to freezes and flooding in the northern hemisphere and a drought in Australia. Wheat futures in September, 2007 for December and March delivery had risen above $9.00 a bushel, prices never seen before. There were complaints in Italy about the high price of pasta.

Production and Consumption

In 2003, global per capita wheat consumption was 67 kg (148 lb), with the highest per capita consumption of 239 kg (527 lb) found in Kyrgyzstan. In 1997, global wheat consumption was 101 kg (223 lb) per capita, with the highest consumption 623 kg (1,373 lb) per capita in Denmark, but most of this (81%) was for animal feed. Wheat is the primary food staple in North Africa and the Middle East, and is growing in popularity in Asia. Unlike rice, wheat production is more widespread globally though China's share is almost one-sixth of the world.

"There is a little increase in yearly crop yield comparison to the year 1990. The reason for this is not in development of sowing area, but the slow and successive increasing of the average yield. Average 2.5 tons wheat was produced on one hectare crop land in the world in the first half of 1990s, however this value was about 3 tons in 2009. In the world per capita wheat producing area continuously decreased between 1990 and 2009 considering the change of world population. There was no significant change in wheat producing area in this period. However, due to the improvement of average yields there is some fluctuation in each year considering the per capita production, but there is no considerable decline. In 1990 per capita production was 111.98 kg/capita/year, while it was already 100.62 kg/capita/year in 2009. The decline is evident and the per capita production level of the year 1990 can not be feasible simultaneously with the growth of world population in spite of the increased average yields. In the whole period the lowest per capita production was in 2006."

In the 20th century, global wheat output expanded by about 5-fold, but until about 1955 most of this reflected increases in wheat crop area, with lesser (about 20%) increases in crop yields per unit area. After 1955 however, there was a dramatic ten-fold increase in the rate of wheat yield improvement per year, and this became the major factor allowing global wheat production to increase. Thus technological innovation and scientific crop management with synthetic nitrogen fertilizer, irrigation and wheat breeding were the main drivers of wheat output growth in the second half of the century. There were some significant decreases in wheat crop area, for instance in North America.

Better seed storage and germination ability (and hence a smaller requirement to retain harvested crop for next year's seed) is another 20th century technological innovation. In Medieval England, farmers saved one-quarter of their wheat harvest as seed for the next crop, leaving only three-quarters for food and feed consumption. By 1999, the global average seed use of wheat was about 6% of output.

Several factors are currently slowing the rate of global expansion of wheat production: population growth rates are falling while wheat yields continue to rise, and the better economic profitability of other crops such as soybeans and maize, linked with investment in modern genetic technologies, has promoted shifts to other crops.

Sugarcane

Sugarcane is any of several species of tall perennial true grasses of the genus *Saccharum*, tribe Andropogoneae, native to the warm temperate to tropical regions of South Asia, and used for sugar production. They have stout jointed fibrous stalks that are rich in sugar, and measure two to six metres (6 to 19 feet) tall. All sugar cane species interbreed and the major commercial cultivars are complex hybrids.

Sugarcane belongs to the grass family (Poaceae), an economically important seed plant family that includes maize, wheat, rice, and sorghum and many forage crops. The main product of sugarcane is sucrose, which accumulates in the stalk internodes. Sucrose, extracted and purified in specialized mill factories, is used as raw material in human food industries or is fermented to produce ethanol. Ethanol is produced on a large scale by the Brazilian sugarcane industry.

Sugarcane is the world's largest crop by production quantity. In 2012, FAO estimates it was cultivated on about 26.0 million hectares, in more than 90 countries, with a worldwide harvest of 1.83 billion tons. Brazil was the largest producer of sugar cane in the world. The next five major producers, in decreasing amounts of production, were India, China, Thailand, Pakistan and Mexico.

The world demand for sugar is the primary driver of sugarcane agriculture. Cane accounts for 80% of sugar produced; most of the rest is made from sugar beets. Sugarcane predominantly grows in the tropical and subtropical regions, and sugar beet predominantly grows in colder temperate regions of the world. Other than sugar, products derived from sugarcane include falernum, molasses, rum, *cachaça* (a traditional spirit from Brazil), bagasse and ethanol. In some regions, people use sugarcane reeds to make pens, mats, screens, and thatch. The young unexpanded inflorescence of *tebu telor* is eaten raw, steamed or toasted, and prepared in various ways in certain island communities of Indonesia.

In India, between the sixth and fourth centuries BC, the Persians, followed by the Greeks, discovered the famous "reeds that produce honey without bees". They adopted and then spread sugar and sugarcane agriculture. A few merchants began to trade in sugar—a luxury and an expensive spice until the 18th century. Before the 18th century, cultivation of sugar cane was largely confined to India. Sugarcane plantations, like cotton farms, were a major driver of large human migrations in the 19th and early 20th century, influencing the ethnic mix, political conflicts and cultural evolution of various Caribbean, South American, Indian Ocean and Pacific island nations.

Cultivation

Sugarcane cultivation requires a tropical or temperate climate, with a minimum of 60 centimetres (24 in) of annual moisture. It is one of the most efficient photosynthesizers in the plant kingdom. It is a C_4 plant, able to convert up to one percent of incident solar energy into biomass. In prime growing regions, such as Mauritius, Dominican Republic, Puerto Rico, India, Indonesia, Pakistan, Peru, Brazil, Bolivia, Colombia, Australia, Ecuador, Cuba, the Philippines, El Salvador and Hawaii, sugarcane crops can produce over 15 kilograms of cane per square metre of sunshine. Once a major crop of the southeastern region of the United States, sugarcane cultivation has declined there in recent decades, and is now primarily confined to Florida and Louisiana.

Sugarcane is cultivated in the tropics and subtropics in areas with a plentiful supply of water, for a continuous period of more than six to seven months each year, either from natural rainfall or through irrigation. The crop does not tolerate severe frosts. Therefore, most of the world's sugarcane is grown between 22°N and 22°S, and some up to 33°N and 33°S. When sugarcane crop is found outside this range, such as the Natal region of South Africa, it is normally due to anomalous climatic conditions in the region, such as warm ocean currents that sweep down the coast. In terms of altitude, sugarcane crop is found up to 1,600 m close to the equator in countries such as Colombia, Ecuador and Peru.

Sugarcane can be grown on many soils ranging from highly fertile well drained mollisols, through heavy cracking vertisols, infertile acid oxisols, peaty histosols to rocky andisols. Both plentiful sunshine and water supplies increase cane production. This has made desert countries with good irrigation facilities such as Egypt as some of the highest yielding sugarcane cultivating regions. .

Although sugarcanes produce seeds, modern stem cutting has become the most common reproduction method. Each cutting must contain at least one bud, and the cuttings are sometimes hand-planted. In more technologically advanced countries like the United States and Australia, billet planting is common. Billets harvested from a mechanical harvester are planted by a machine that opens and recloses the ground. Once planted, a stand can be harvested several times; after each harvest, the cane sends up new stalks, called ratoons. Successive harvests give decreasing yields, eventually justifying replanting. Two to 10 harvests are usually made depending on the type of culture. In a country with a mechanical agriculture looking for a high production of large fields, like in North America, sugar canes are replanted after two or three harvests to avoid a lowering in yields. In countries with a more traditional type of agriculture with smaller fields and hand harvesting, like in the French island la Réunion, sugar canes are often harvested up to 10 years before replanting.

Sugarcane is harvested by hand and mechanically. Hand harvesting accounts for more than half of production, and is dominant in the developing world. In hand harvesting, the field is first set on fire. The fire burns dry leaves, and chases away or kills any lurking venomous snakes, without harming the stalks and roots. Harvesters then cut the cane just above ground-level using cane knives or machetes. A skilled harvester can cut 500 kilograms (1,100 lb) of sugarcane per hour.

Mechanical harvesting uses a combine, or sugarcane harvester. The Austoft 7000 series, the original modern harvester design, has now been copied by other companies, including Cameco / John Deere. The machine cuts the cane at the base of the stalk, strips the leaves, chops the cane into consistent lengths and deposits it into a transporter following alongside. The harvester then blows the trash back onto the field. Such machines can harvest 100 long tons (100 t) each hour; however, harvested cane must be rapidly processed. Once cut, sugarcane begins to lose its sugar content, and damage to the cane during mechanical harvesting accelerates this decline. This decline is offset because a modern chopper harvester can complete the harvest faster and more efficiently than hand cutting and loading. Austoft also developed a series of hydraulic high-lift infield transporters to work alongside their harvesters to allow even more rapid transfer of cane to, for example, the nearest railway siding. This mechanical harvesting doesn't require the field to be set on fire; the remains left in the field by the machine consist of the top of the sugar cane and the dead leaves, which act as mulch for the next round of planting.

Processing

Traditionally, sugarcane processing requires two stages. Mills extract raw sugar from freshly harvested cane and "mill-white" sugar is sometimes produced immediately after the first stage at sugar-extraction mills, intended for local consumption. Sugar crystals appear naturally in white colour during the crystallization process. Sulfur dioxide is added to inhibit the formation of colour-inducing molecules as well as to stabilize the sugar juices during evaporation. Refineries, often located nearer to consumers in North America, Europe, and Japan, then produce refined white sugar, which is 99 percent sucrose. These two stages are slowly merging. Increasing affluence in the sugar-producing tropics increased demand for refined sugar products, driving a trend toward combined milling and refining.

Milling

Sugarcane processing produces cane sugar (sucrose) from sugarcane. Other products of the processing include bagasse, molasses, and filtercake.

Bagasse, the residual dry fibre of the cane after cane juice has been extracted, is used for several purposes:

- fuel for the boilers and kilns,
- production of paper, paperboard products and reconstituted panelboard,
- agricultural mulch, and
- as a raw material for production of chemicals.

Figure: *Santa Elisa sugarcane processing plant in Sertɀozinho, one of the largest and oldest in Brazil*

The primary use of bagasse and bagasse residue is as a fuel source for the boilers in the generation of process steam in sugar plants. Dried filtercake is used as an animal feed supplement, fertilizer, and source of sugarcane wax.

Molasses is produced in two forms: Blackstrap, which has a characteristic strong flavour because of its vitamin and mineral content, and a purer molasses syrup. Blackstrap molasses is sold as a food and dietary supplement. It is also a common ingredient in animal feed, is used to produce ethanol and rum, and in the manufacturing of citric acid. Purer molasses syrups are sold as molasses, and may also be blended with maple syrup, invert sugars, or corn syrup. Both forms of molasses are used in baking.

Refining

Sugar refining further purifies the raw sugar. It is first mixed with heavy syrup and then centrifuged in a process called "affination". Its purpose is to wash away the sugar crystals' outer coating, which is less pure than the crystal interior. The remaining sugar is then dissolved to make a syrup, about 60 percent solids by weight.

The sugar solution is clarified by the addition of phosphoric acid and calcium hydroxide, which combine to precipitate calcium phosphate. The calcium phosphate particles entrap some impurities and absorb others, and then float to the top of the tank, where they can be skimmed off. An alternative to this "phosphatation" technique is "carbonatation", which is similar, but uses carbon dioxide and calcium hydroxide to produce a calcium carbonate precipitate.

After filtering any remaining solids, the clarified syrup is decolorized by filtration through activated carbon. Bone char or coal-based activated carbon is traditionally used in this role. Some remaining colour-forming impurities adsorb to the carbon. The purified syrup is then concentrated to supersaturation and repeatedly crystallized in a vacuum, to produce white refined sugar. As in a sugar mill, the sugar crystals are separated from the molasses by centrifuging. Additional sugar is recovered by blending the remaining syrup with the washings from affination and again crystallizing to produce brown sugar. When no more sugar can be economically recovered, the final molasses still contains 20–30 percent sucrose and 15–25 percent glucose and fructose.

To produce granulated sugar, in which individual grains do not clump, sugar must be dried, first by heating in a rotary dryer, and then by blowing cool air through it for several days.

Ribbon Cane Syrup

Ribbon cane is a subtropical type that was once widely grown in the southern United States, as far north as coastal North Carolina. The juice was extracted with horse or mule-powered crushers; the juice was boiled, like maple syrup, in a flat pan, and then used in the syrup form as a food sweetener. It is not currently a commercial crop, but a few growers find ready sales for their product.

Pollution from Sugarcane Processing

Particulate matter, combustion products, and volatile organic compounds are the primary pollutants emitted during the sugarcane processing. Combustion products include nitrogen oxides (NO_X), carbon monoxide (CO), CO_2, and sulfur oxides (SO_X). Potential emission sources include the sugar granulators, sugar conveying and packaging equipment, bulk loadout operations, boilers, granular carbon and char regeneration kilns, regenerated adsorbent transport systems, kilns and handling equipment (at some facilities), carbonation tanks, multi-effect evaporator stations, and vacuum boiling pans. Modern pollution prevention technologies are capable of addressing all of these potential pollutants.

Production

Brazil led the world in sugarcane production in 2011 with a 734 000 TMT harvest. India was the second largest producer with 342 382 TMT tons, and China the third largest producer with 115 125 TMT tons harvest. The average worldwide yield of sugarcane crops in 2011 was 70.54 tons per hectare. The most productive farms in the world were in Ethiopia with a nationwide average sugarcane crop yield of 126.93 tons per hectare.

The theoretical possible yield for sugar cane, according to 1983 study of Duke, is about 280 metric tons per hectare per year, and small experimental plots in Brazil have demonstrated yields of 236–280 metric tons of fresh cane per hectare. The most promising region for high yield sugarcane production were in sun drenched, irrigated farms of northern Africa, and other deserts with plentiful water from river or irrigation canals.

In the United States, sugarcane is grown commercially in Florida, Hawaii, Louisiana, and Texas.

Brazil uses sugarcane to produce sugar and ethanol for gasoline-ethanol blends (gasohol), a locally popular transportation fuel. In India, sugarcane is used to produce sugar, jaggery and alcoholic beverages.

Cane Ethanol

Ethanol is generally available as a byproduct of sugar production. It can be used as a biofuel alternative to gasoline, and is widely used in cars in Brazil. It is an alternative to gasoline, and may become the primary product of sugarcane processing, rather than sugar.

In Brazil, gasoline is required to contain at least 22 percent bioethanol. This bioethanol is sourced from Brazil's large sugarcane crop.

The production of ethanol from sugar cane is more energy efficient than from corn or sugar beets or palm/vegetable oils, particularly if cane bagasse is used to produce heat and power for the process. Furthermore, if biofuels are used for crop production and transport, the fossil energy input needed for each ethanol energy unit can be very low. EIA estimates that with an integrated sugar cane to ethanol technology, the well-to-wheels CO_2 emissions can be 90 percent lower than conventional gasoline.

A Textbook on Renewable Energy Describes the Energy Transformation: Presently, 75 tons of raw sugar cane are produced annually per hectare in Brazil. The cane delivered to the processing plant is called burned and cropped (b&c), and represents 77% of the mass of the raw cane. The reason for this reduction is that the stalks are separated from the leaves (which are burned and whose ashes are left in the field as fertilizer), and from the roots that remain in the ground to sprout for the next crop. Average cane production is, therefore, 58 tons of b&c per hectare per year.

Each ton of b&c yields 740 kg of juice (135 kg of sucrose and 605 kg of water) and 260 kg of moist bagasse (130 kg of dry bagasse). Since the higher heating value of sucrose is 16.5 MJ/kg, and that of the bagasse is 19.2 MJ/kg, the total heating value of a ton of b&c is 4.7 GJ of which 2.2 GJ come from the sucrose and 2.5 from the bagasse.

Per hectare per year, the biomass produced corresponds to 0.27 TJ. This is equivalent to 0.86 W per square metre. Assuming an average insolation of 225 W per square metre, the photosynthetic efficiency of sugar cane is 0.38%.

The 135 kg of sucrose found in 1 ton of b&c are transformed into 70 litres of ethanol with a combustion energy of 1.7 GJ. The practical sucrose-ethanol conversion efficiency is, therefore, 76% (compare with the theoretical 97%).

One hectare of sugar cane yields 4,000 litres of ethanol per year (without any additional energy input, because the bagasse produced exceeds the amount needed to distill the final product). This, however, does not include the energy used in tilling, transportation, and so on. Thus, the solar energy-to-ethanol conversion efficiency is 0.13%.

Bagasse Applications

Sugarcane is a major crop in many countries. It is one of the plants with the highest bioconversion efficiency. Sugarcane crop is able to efficiently fix solar energy, yielding some 55 tonnes of dry matter per hectare of land annually. After harvest, the crop produces sugar juice and bagasse, the fibrous dry matter. This dry matter is biomass with potential as fuel for energy production.

Sugarcane bagasse is a potentially abundant source of energy for large producers of sugarcane, such as Brazil, India and China. According to one report, with use of latest technologies, bagasse produced annually in Brazil has the potential of meeting 20 percent of Brazil's energy consumption by 2020.

Electricity Production

A number of countries, in particular those devoid of any fossil fuel, have implemented energy conservation and efficiency measures to minimize energy utilized in cane processing and furthermore export any excess electricity to the grid. Bagasse is usually burned to produce steam, which in turn creates electricity. Current technologies, such as those in use in Mauritius, produce over 100 KWh of electricity per tonne of bagasse. With a total world harvest of over 1 billion tonnes of sugar cane per year, the global energy potential from bagasse is over 100,000 GWh. Using Mauritius as a reference, an annual potential of 10,000 GWh of additional electricity could be produced throughout Africa. Electrical generation from bagasse could become quite important, particularly to the rural populations of sugarcane producing nations.

Recent cogeneration technology plants are being designed to produce from 200 to over 300 KWh of electricity per tonne of bagasse. As sugarcane is a seasonal crop, shortly after harvest the supply of bagasse would peak, requiring power generation plants to strategically manage the storage of bagasse.

Biogas Production

A greener alternative to burning bagasse for the production of electricity is to convert bagasse into biogas. Technologies are being developed to use enzymes to transform bagasse into advanced biofuel and biogas. Not only could this process realize a greater energy potential, the release of greenhouse gases would be drastically less than simply burning bagasse.

Sugarcane as Food

In most countries where sugarcane is cultivated, there are several foods and popular dishes derived directly from it, such as:

- Raw sugarcane: chewed to extract the juice
- *Sayur nganten*: an Indonesian soup made with the stem of trubuk (*Saccharum edule*), a type of sugarcane.
- Sugarcane juice: a combination of fresh juice, extracted by hand or small mills, with a touch of lemon and ice to make a popular drink, known variously as *air tebu, usacha rass, guarab, guarapa, guarapo, papelón, aseer asab, ganna sharbat, mosto, caldo de cana.*
- Syrup: a traditional sweetener in soft drinks, now largely supplanted in the US by high fructose corn syrup, which is less expensive because of corn subsidies and sugar tariffs.
- Molasses: used as a sweetener and a syrup accompanying other foods, such as cheese or cookies
- Jaggery: a solidified molasses, known as *gur* or *gud* or *gul* in India, is traditionally produced by evaporating juice to make a thick sludge, and then cooling and molding it in buckets. Modern production partially freeze dries the juice to reduce caramelization and lighten its colour. It is used as sweetener in cooking traditional entrees, sweets and desserts.
- Falernum: a sweet, and lightly alcoholic drink made from sugarcane juice
- *Cachaça*: the most popular distilled alcoholic beverage in Brazil; a liquor made of the distillation of sugarcane juice.
- Rum: is a liquor made from sugarcane products, typically molasses but sometimes also cane juice. It is most commonly produced in the Caribbean and environs.
- Basi: is a fermented alcoholic beverage made from sugarcane juice produced in the Philippines and Guyana.
- *Panela*: solid pieces of sucrose and fructose obtained from the boiling and evaporation of sugarcane juice; a food staple in Colombia and other countries in South and Central America
- *Rapadura*: a sweet flour that is one of the simplest refinings of sugarcane juice, common in Latin American countries such as Brazil, Argentina and Venezuela (where it is known as *papelón*) and the Caribbean.

- Rock candy: crystallized cane juice
- Gβteau de Sirop

Natural Rubber

Figure: *Latex being collected from a tapped rubber tree, Cameroon*

Natural rubber, also called India rubber or *caoutchouc*, as initially produced, consists of polymers of the organic compound isoprene, with minor impurities of other organic compounds plus water. Forms of polyisoprene that are used as natural rubbers are classified as elastomers. Currently, rubber is harvested mainly in the form of the latex from certain trees. The latex is a sticky, milky colloid drawn off by making incisions into the bark and collecting the fluid in vessels in a process called "tapping". The latex then is refined into rubber ready for commercial processing. Natural rubber is used extensively in many applications and products, either alone or in combination with other materials. In most of its useful forms, it has a large stretch ratio and high resilience, and is extremely waterproof.

Varieties

The major commercial source of natural rubber latex is the Parα rubber tree (*Hevea brasiliensis*), a member of the spurge family,

Euphorbiaceae. This species is widely used because it grows well under cultivation and a properly managed tree responds to wounding by producing more latex for several years.

Congo rubber, formerly a major source of rubber, came from vines in the genus *Landolphia* (*L. kirkii*, *L. heudelotis*, and *L. owariensis*). These cannot be cultivated, and the intense drive to collect latex from wild plants was responsible for many of the atrocities committed under the Congo Free State.

Many other plants produce forms of latex rich in isoprene polymers, though not all produce usable forms of polymer as easily as the Pará rubber tree does; some of them require more elaborate processing to produce anything like usable rubber, and most are more difficult to tap. Some produce other desirable materials, for example gutta-percha (*Palaquium gutta*) and chicle from *Manilkara* species. Others that have been commercially exploited, or at least have shown promise as sources of rubber, include the rubber fig (*Ficus elastica*), Panama rubber tree (*Castilla elastica*), various spurges (*Euphorbia* spp.), lettuce (*Lactuca* species), the related *Scorzonera tau-saghyz*, various *Taraxacum* species, including common dandelion (*Taraxacum officinale*) and Russian dandelion (*Taraxacum kok-saghyz*), and guayule (*Parthenium argentatum*). The term gum rubber is sometimes applied to the tree-obtained version of natural rubber in order to distinguish it from the synthetic version.

Discovery of Commercial Potential

The Para rubber tree is indigenous to South America. Charles Marie de La Condamine is credited with introducing samples of rubber to the *Académie Royale des Sciences* of France in 1736. In 1751, he presented a paper by François Fresneau to the Académie (eventually published in 1755) which described many of the properties of rubber. This has been referred to as the first scientific paper on rubber. In England, Joseph Priestley, in 1770, observed that a piece of the material was extremely good for rubbing off pencil marks on paper, hence the name "rubber". Later, it slowly made its way around England.

South America remained the main source of the limited amounts of latex rubber used during much of the 19th century. The trade was well protected and exporting seeds from Brazil was said to be a capital offence, although there was no law against it. Nevertheless, in 1876, Henry Wickham smuggled 70,000 Para rubber tree seeds from Brazil and delivered them to Kew Gardens, England. Only 2,400 of these germinated after which the seedlings were then sent to India, Ceylon

(Sri Lanka), Indonesia, Singapore, and British Malaya. Malaya (now Peninsular Malaysia) was later to become the biggest producer of rubber. In the early 1900s, the Congo Free State in Africa was also a significant source of natural rubber latex, mostly gathered by forced labour. Liberia and Nigeria also started production of rubber.

In India, commercial cultivation of natural rubber was introduced by the British planters, although the experimental efforts to grow rubber on a commercial scale in India were initiated as early as 1873 at the Botanical Gardens, Calcutta. The first commercial *Hevea* plantations in India were established at Thattekadu in Kerala in 1902.

In Singapore and Malaya, commercial production of rubber was heavily promoted by Sir Henry Nicholas Ridley, who served as the first Scientific Director of the Singapore Botanic Gardens from 1888 to 1911. He distributed rubber seeds to many planters and developed the first technique for tapping trees for latex without causing serious harm to the tree. Because of his very fervent promotion of this crop, he is popularly remembered by the nickname "Mad Ridley".

Properties

Rubber exhibits unique physical and chemical properties. Rubber's stress-strain behaviour exhibits the Mullins effect and the Payne effect, and is often modelled as hyperelastic. Rubber strain crystallizes.

Due to the presence of a double bond in each repeat unit, natural rubber is susceptible to vulcanisation and sensitive to ozone cracking.

The two main solvents for rubber are turpentine and naphtha (petroleum). The former has been in use since 1764 when François Fresnau made the discovery. Giovanni Fabbroni is credited with the discovery of naphtha as a rubber solvent in 1779. Because rubber does not dissolve easily, the material is finely divided by shredding prior to its immersion.

An ammonia solution can be used to prevent the coagulation of raw latex while it is being transported from its collection site.

Elasticity

On a microscopic scale, relaxed rubber is a disorganized cluster of erratically changing wrinkled chains. In stretched rubber, the chains are almost linear. The restoring force is due to the preponderance of wrinkled conformations over more linear ones.

Cooling below the glass transition temperature still permits local conformational changes but a reordering is practically impossible because of the larger energy barrier for the concerted movement of longer chains. "Frozen" rubber's elasticity is low and strain results from small changes of bond lengths and angles. (This caused the Challenger disaster, where flattened o-rings failed to relax to fill a widening gap.) The glass transition is fast and reversible: the force resumes on heating.

The parallel chains of stretched rubber are susceptible to crystallization. This takes some time because turns of twisted chains have to move out of the way of the growing crystallites. Crystallization has occurred, for example, when, after days, an inflated toy balloon is found withered at a relatively large remaining volume. Where it is touched, it shrinks because the temperature of the hand is enough to melt the crystals.

Vulcanization of rubber creates disulfide bonds between chains, which limits the degrees of freedom and results in chains that tighten more quickly for a given strain, thereby increasing the elastic force constant and making the rubber harder and less extensible.

Chemical Makeup

Latex is the polymer cis-1,4-polyisoprene – with a molecular weight of 100,000 to 1,000,000 daltons. Typically, a small percentage (up to 5% of dry mass) of other materials, such as proteins, fatty acids, resins, and inorganic materials (salts) are found in natural rubber. Polyisoprene can also be created synthetically, producing what is sometimes referred to as "synthetic natural rubber", but the synthetic and natural routes are completely different.

Some natural rubber sources, such as gutta-percha, are composed of trans-1,4-polyisoprene, a structural isomer that has similar, but not identical, properties.

Natural rubber is an elastomer and a thermoplastic. Once the rubber is vulcanized, it will turn into a thermoset. Most rubber in everyday use is vulcanized to a point where it shares properties of both; i.e., if it is heated and cooled, it is degraded but not destroyed.

The final properties of a rubber item depend not just on the polymer, but also on modifiers and fillers, such as carbon black, factice, whiting, and a host of others.

Biosynthesis

Rubber particles are formed in the cytoplasm of specialized latex-producing cells called laticifers within rubber plants. Rubber particles are surrounded by a single phospholipid membrane with hydrophobic tails pointed inward. The membrane allows biosynthetic proteins to be sequestered at the surface of the growing rubber particle, which allows new monomeric units to be added from outside the biomembrane, but within the lacticifer. The rubber particle is an enzymatically active entity that contains three layers of material, the rubber particle, a biomembrane, and free monomeric units.

The biomembrane is held tightly to the rubber core due to the high negative charge along the double bonds of the rubber polymer backbone. Free monomeric units and conjugated proteins make up the outer layer. The rubber precursor is isopentenyl pyrophosphate (an allylic compound), which elongates by Mg^{2+}-dependent condensation by the action of rubber transferase.

The monomer adds to the pyrophosphate end of the growing polymer. The process displaces the terminal high-energy pyrophosphate. The reaction produces a cis polymer. The initiation step is catalyzed by prenyltransferase, which converts three monomers of isopentenyl pyrophosphate into farnesyl pyrophosphate. The farnesyl pyrophosphate can bind to rubber transferase to elongate a new rubber polymer.

The required isopentenyl pyrophosphate is obtained from the mevalonate pathway, which is derives from acetyl-CoA in the cytosol. In plants, isoprene pyrophosphate can also be obtained from 1-deox-D-xyulose-5-phosphate/2-C-methyl-D-erythritol-4-phosphate pathway within plasmids. The relative ratio of the farnesyl pyrophosphate initiator unit and isoprenyl pyrophosphate elongation monomer determines the rate of new particle synthesis versus elongation of existing particles. Though rubber is known to be produced by only one enzyme, extracts of latex have shown numerous small molecular weight proteins with unknown function. The proteins possibly serve as cofactors, as the synthetic rate decreases with complete removal.

Current Sources

Close to 21 million tons of rubber were produced in 2005, of which approximately 42% was natural. Since the bulk of the rubber produced is of the synthetic variety, which is derived from petroleum, the price of natural rubber is determined, to a large extent, by the prevailing global price of crude oil. Today, Asia is the main source of natural rubber, accounting for about 94% of output in 2005. The three largest producing countries, Thailand, Indonesia (2.4m tons) and Malaysia, together account for around 72% of all natural rubber production. Natural rubber is not cultivated widely in its native continent of South America due to the existence of South American leaf blight, and other natural predators of the rubber tree.

Cultivation

Rubber latex is extracted from rubber trees. The economic life period of rubber trees in plantations is around 32 years – up to 7 years of immature phase and about 25 years of productive phase.

The soil requirement of the plant is generally well-drained, weathered soil consisting of laterite, lateritic types, sedimentary types, nonlateritic red, or alluvial soils.

The climatic conditions for optimum growth of rubber trees are:

- Rainfall of around 250 cm evenly distributed without any marked dry season and with at least 100 rainy days per year
- Temperature range of about 20 to 34°C, with a monthly mean of 25 to 28°C
- High atmospheric humidity of around 80%
- Bright sunshine amounting to about 2000 hours per year at the rate of six hours per day throughout the year
- Absence of strong winds

Many high-yielding clones have been developed for commercial planting. These clones yield more than 2,000 kg of dry rubber per hectare per year, when grown under ideal conditions.

Collection

In places such as Kerala, where coconuts are in abundance, the half shell of coconut is used as the collection container for the latex, but glazed pottery or aluminium or plastic cups are more common elsewhere. The cups are supported by a wire that encircles the tree. This wire incorporates a spring so it can stretch as the tree grows. The latex is led into the cup by a galvanised "spout" knocked into the bark. Tapping normally takes place early in the morning, when the internal pressure of the tree is highest.

A good tapper can tap a tree every 20 seconds on a standard half-spiral system, and a common daily "task" size is between 450 and 650 trees. Trees are usually tapped on alternate or third days, although many variations in timing, length, and number of cuts are used. The latex, which contains 25–40% dry rubber, is in the bark, so the tapper must avoid cutting right through to the wood, else the growing cambial layer will be damaged and the renewing bark will be badly deformed, making later tapping difficult. It is usual to tap a pannel at least twice, sometimes three times, during the tree's life. The economic life of the tree depends on how well the tapping is carried out, as the critical factor is bark consumption.

A standard in Malaysia for alternate daily tapping is 25 cm (vertical) bark consumption per year. The latex tubes in the bark ascend in a spiral to the right. For this reason, tapping cuts usually ascend to the left to cut more tubes.

The trees will drip latex for about four hours, stopping as latex coagulates naturally on the tapping cut, thus blocking the latex tubes in the bark. Tappers usually rest and have a meal after finishing their tapping work, then start collecting the liquid "field latex" at about midday. Some trees will continue to drip after the collection and this leads to a small amount of "cup lump" which is collected at the next tapping. The latex that coagulates on the cut is also collected as "tree lace". Tree lace and cup lump together account for 10–20% of the dry rubber produced. Latex that drips onto the ground, "earth scrap", is also collected periodically for processing of low-grade product.

Field Coagula

There are four types of field coagula, "cuplump", "treelace", "smallholders' lump" and "earth scrap". Each has significantly different properties.

Cuplump is the coagulated material found in the collection cup when the tapper next visits the tree to tap it again. It arises from latex clinging to the walls of the cup after the latex was last poured into the bucket, and from late-dripping latex exuded before the latex-carrying vessels of the tree become blocked. It is of higher purity and of greater value than the other three types.

Treelace is the coagulum strip that the tapper peels off the previous cut before making a new cut. It usually has higher copper and manganese contents than cuplump. Both copper and manganese are pro-oxidants and can lower the physical properties of the dry rubber.

Smallholders' lump is produced by smallholders who collect rubber from trees a long way away from the nearest factory. Many Indonesian smallholders, who grow paddy in remote areas, tap dispersed trees on their way to work in the paddy fields and collect the latex (or the coagulated latex) on their way home.

As it is often impossible to preserve the latex sufficiently to get it to a factory that processes latex in time for it to be used to make high quality products, and as the latex would anyway have coagulated by the time it reached the factory, the smallholder will coagulate it by any means available, in any container available. Some smallholders use small containers, buckets etc., but often the latex is coagulated in holes in the ground, which are usually (but not always) lined with plastic. Acidic materials and fermented fruit juices are used to coagulate the latex – a form of assisted biological coagulation. Little care is taken to exclude twigs, leaves, and even bark from the lumps that are formed, which may also include treelace collected by the smallholder.

Earth scrap is the material that gathers around the base of the tree. It arises from latex overflowing from the cut and running down the bark of the tree, from rain flooding a collection cup containing latex, and from spillage from tappers' buckets during collection. It contains soil and other contaminants, and has variable rubber content depending on the amount of contaminants mixed with it. Earth scrap is collected by the field workers two or three times a year and may be cleaned in a scrap-washer to recover the rubber, or sold off to a contractor who will clean it and recover the rubber. It is of very low quality and under no circumstances should it be included in block rubber or brown crepe.

Processing

The latex will coagulate in the cups if kept for long. The latex has to be collected before coagulation. The collected latex, "field latex", is transferred into coagulation tanks for the preparation of dry rubber or transferred into air-tight containers with sieving for ammoniation. Ammoniation is necessary to preserve the latex in a colloidal state for longer periods of time.

Latex is generally processed into either latex concentrate for manufacture of dipped goods or it can be coagulated under controlled, clean conditions using formic acid. The coagulated latex can then be processed into the higher-grade, technically specified block rubbers such as SVR 3L or SVR CV or used to produce Ribbed Smoke Sheet grades.

Naturally coagulated rubber (cup lump) is used in the manufacture of TSR10 and TSR20 grade rubbers. The processing of the rubber for these grades is a size reduction and cleaning process to remove contamination and prepare the material for the final stage of drying.

The dried material is then baled and palletized for storage and shipment in various methods of transportation.

Transportation

Natural rubber latex is shipped from factories in south-west Asia, South America, and North Africa to destinations around the world. As the cost of natural rubber has risen significantly, the shipping methods which offer the lowest cost per unit of weight are preferred. Depending on the destination, warehouse availability, and transportation conditions, some methods are more suitable to certain buyers than others. In international trade, latex rubber is mostly shipped in 20-foot ocean containers. Inside the ocean container, various types of smaller containers are used by factories to store latex rubber.

Uses

Contemporary Manufacturing

Around 25 million tonnes of rubber is produced each year, of which 42 percent is natural rubber. The remainder is synthetic rubber derived from petrochemical sources. Around 70 percent of the world's natural rubber is used in tires. The top end of latex production results in latex products such as surgeons' gloves, condoms, balloons and other relatively high-value products. The mid-range which comes from the technically-specified natural rubber materials ends up largely in tires but also in conveyor belts, marine products, windshield wipers and miscellaneous rubber goods. Natural rubber offers good elasticity, while synthetic materials tend to offer better resistance to environmental factors such as oils, temperature, chemicals or ultraviolet light and suchlike. "Cured rubber" is rubber which has been compounded and subjected to the vulcanisation process which creates cross-links within the rubber matrix.

Prehistoric Uses

The first use of rubber was by the Olmecs, who centuries later passed on the knowledge of natural latex from the *Hevea* tree in 1600 BC to the ancient Mayans. They boiled the harvested latex to make a ball for a Mesoamerican ballgame.

Pre-World War II Manufacturing

Other significant uses of rubber are door and window profiles, hoses, belts, gaskets, matting, flooring, and dampeners (antivibration mounts) for the automotive industry. Gloves (medical, household and industrial) and toy balloons are also large consumers of rubber, although the type of rubber used is concentrated latex. Significant tonnage of rubber is used as adhesives in many manufacturing industries and products, although the two most noticeable are the paper and the carpet industries. Rubber is also commonly used to make rubber bands and pencil erasers.

Pre-World War II textile Applications

Rubber produced as a fibre, sometimes called 'elastic', has significant value for use in the textile industry because of its excellent elongation and recovery properties. For these purposes, manufactured rubber fibre is made as either an extruded round fibre or rectangular fibres that are cut into strips from extruded film. Because of its low dye acceptance, feel and appearance, the rubber fibre is either covered by yarn of another fibre or directly woven with other yarns into the fabric. In the early 1900s, for example, rubber yarns were used in foundation garments.

While rubber is still used in textile manufacturing, its low tenacity limits its use in lightweight garments because latex lacks resistance to oxidizing agents and is damaged by aging, sunlight, oil, and perspiration. Seeking a way to address these shortcomings, the textile industry has turned to neoprene (polymer of chloroprene), a type of synthetic rubber, as well as another more commonly used elastomer fibre, spandex (also known as elastane), because of their superiority to rubber in both strength and durability.

Vulcanization

Natural rubber is often vulcanized, a process by which the rubber is heated and sulfur, peroxide or bisphenol are added to improve resistance and elasticity, and to prevent it from perishing. The development of vulcanization is most closely associated with Charles Goodyear in 1839. Before World War II era manufacturing, carbon black was often used as an additive to rubber to improve its strength, especially in vehicle tires.

Allergic Reactions

Some people have a serious latex allergy, and exposure to natural latex rubber products such as latex gloves can cause anaphylactic shock. The antigenic proteins found in *Hevea* latex may be deliberately reduced (though not eliminated) through processing.

Latex from non-*Hevea* sources, such as Guayule, can be used without allergic reaction by persons with an allergy to *Hevea* latex.

Some allergic reactions are not to the latex itself, but from residues of chemicals used to accelerate the cross-linking process. Although this may be confused with an allergy to latex, it is distinct from it, typically taking the form of Type IV hypersensitivity in the presence of traces of specific processing chemicals.

Alternative Sources

Dandelion milk had been known to contain latex for a long time. The latex exhibited the same quality as the natural rubber from rubber trees. Yet in the wild types of dandelion, the latex content is low and varies greatly.

In Nazi Germany, research projects tried to use dandelions as a base for rubber production, but failed.

2013 by inhibiting one key enzyme and using modern cultivation methods and optimization techniques, scientists in the Fraunhofer Institute for Molecular Biology and Applied Ecology (IME) in Germany developed a cultivar that is suitable for commercial production of natural rubber. In collaboration with Continental Tires, IME is building a pilot facility. The first prototype test tires made with blends from dandelion-rubber are scheduled to be tested on public roads over the next few years.

Microbial Degradation

Natural rubber is susceptible to degradation by a wide range of bacteria.

Coffee

Coffee is a brewed beverage prepared from the roasted or baked seeds of several species of an evergreen shrub of the genus *Coffea.* The two most common sources of coffee beans are the highly regarded *Coffea arabica,* and the "robusta" form of the hardier *Coffea canephora.* The latter is resistant to the coffee leaf rust (*Hemileia vastatrix*), but has a more bitter taste. Coffee plants are cultivated in more than 70 countries, primarily in equatorial Latin America, Southeast Asia, and Africa. Once ripe, coffee "berries" are picked, processed and dried to yield the seeds inside. The seeds are then roasted to varying degrees, depending on the desired flavour, before being ground and brewed to create coffee.

Coffee is slightly acidic (pH 5.0–5.1) and can have a stimulating effect on humans because of its caffeine content. It is one of the most popular drinks in the world. It can be prepared and presented in a variety of ways. The effect of coffee on human health has been a subject of many studies; however, results have varied in terms of coffee's relative benefit. The majority of recent research suggests that moderate coffee consumption is benign or mildly beneficial in healthy adults. However, the diterpenes in coffee may increase the risk of heart disease.

Coffee cultivation first took place in Abyssinia the earliest credible evidence of coffee-drinking appears in the middle of the 15th century in the Sufi shrines of Yemen.

In the Horn of Africa and Yemen, coffee was used in local religious ceremonies. As these ceremonies conflicted with the beliefs of the Christian church, the Ethiopian Church banned the secular consumption of coffee until the reign of Emperor Menelik II. The beverage was also banned in Ottoman Turkey during the 17th century for political reasons, and was associated with rebellious political activities in Europe.

An important export commodity, coffee was the top agricultural export for twelve countries in 2004, and it was the world's seventh-largest legal agricultural export by value in 2005. Green (unroasted) coffee is one of the most traded agricultural commodities in the world. Some controversy is associated with coffee cultivation and its impact on the environment. Consequently, organic coffee is an expanding market.

Cultivation

The traditional method of planting coffee is to place 20 seeds in each hole at the beginning of the rainy season. This method loses about 50% of the seeds' potential, as about half fail to sprout. A more effective method of growing coffee, used in Brazil, is to raise seedlings in nurseries that are then planted outside at six to twelve months. Coffee is often intercropped with food crops, such as corn, beans, or rice during the first few years of cultivation as farmers become familiar with its requirements.

Of the two main species grown, arabica coffee (from *C. arabica*) is generally more highly regarded than robusta coffee (from *C. canephora*); robusta tends to be bitter and have less flavour but better body than arabica. For these reasons, about three-quarters of coffee cultivated worldwide is *C. arabica*. Robusta strains also contain about 40–50% more caffeine than arabica. Consequently, this species is used as an inexpensive substitute for arabica in many commercial coffee blends. Good quality robusta seeds are used in traditional Italian espresso blends to provide a full-bodied taste and a better foam head (known as *crema*).

However, *Coffea canephora* is less susceptible to disease than *C. arabica* and can be cultivated in lower altitudes and warmer climates where *C. arabica* will not thrive. The robusta strain was first collected in 1890 from the Lomani River, a tributary of the Congo River, and was conveyed from Zaire (now the Democratic Republic of Congo) to Brussels to Java around 1900. From Java, further breeding resulted in the establishment of robusta plantations in many countries. In particular, the spread of the devastating coffee leaf rust (*Hemileia vastatrix*), to which *C. arabica* is vulnerable, hastened the uptake of the resistant robusta. Coffee leaf rust is found in virtually all countries that produce coffee.

Over 900 species of insect have been recorded as pests of coffee crops worldwide. Of these, over a third are beetles, and over a quarter are bugs. Some 20 species of nematodes, 9 species of mites, several snails and slugs also attack the crop. Birds and rodents sometimes eat coffee berries but their impact is minor compared to invertebrates. In general, *arabica* is the more sensitive species to invertebrate predation overall. Each part of the coffee plant is assailed by different animals. Nematodes attack the roots, and borer beetles burrow into stems and woody material, the foliage is attacked by over 100 species of larvae (caterpillars) of butterflies and moths.

Mass spraying of insecticides has often proven disastrous, as the predators of the pests are more sensitive than the pests themselves. Instead, integrated pest management has developed, using techniques such as targeted treatment of pest outbreaks, and managing crop environment away from conditions favouring pests. Branches infested with scale are often cut and left on the ground, which promotes scale parasites to not only attack the scale on the fallen branches but in the plant as well.

The 2-mm-long coffee berry borer beetle (*Hypothenemus hampei*) is the most damaging insect pest to the world's coffee industry, destroying up to 50 percent or more of the coffee berries on plantations in most coffee-producing countries.

The adult female beetle nibbles a single tiny hole in a coffee berry and lays 35 to 50 eggs. Inside, the offspring grow, mate, and then emerge from the commercially ruined berry to disperse, repeating the cycle. Pesticides are mostly ineffective because the beetle juveniles are protected inside the berry nurseries, but they are vulnerable to predation by birds when they emerge.

When groves of trees are nearby, the American Yellow Warbler, Rufous-capped Warbler and other insectivorous birds have been shown to reduce by 50 percent the number of coffee berry borer beetles in Costa Rica coffee plantations.

World Production

In 2011 Brazil was the world leader in production of green coffee, followed by Vietnam, Indonesia and Colombia. Arabica coffee seeds are cultivated in Latin America, eastern Africa, Arabia, or Asia. Robusta coffee seeds are grown in western and central Africa, throughout southeast Asia, and to some extent in Brazil.

Seeds from different countries or regions can usually be distinguished by differences in flavour, aroma, body, and acidity. These taste characteristics are dependent not only on the coffee's growing region, but also on genetic subspecies (varietals) and processing. Varietals are generally known by the region in which they are grown, such as Colombian, Java and Kona.

Ecological Effects

Originally, coffee farming was done in the shade of trees that provided a habitat for many animals and insects. Remnant forest trees were used for this purpose, but many species have been planted as well. These include leguminous trees of the genera *Acacia*, *Albizia*, *Cassia*, *Erythrina*, *Gliricidia*, *Inga*, and *Leucaena*, as well as the nitrogen-fixing non-legume sheoaks of the genus *Casuarina*, and the silky oak *Grevillea robusta*.

This method commonly referred to as the traditional shaded method, or "shade-grown". Starting in the 1970s, many farmers switched their production method to sun cultivation, in which coffee is grown in rows under full sun with little or no forest canopy. This causes berries to ripen more rapidly and bushes to produce higher yields, but requires the clearing of trees and increased use of fertilizer and pesticides, which damage the environment and cause health problems.

Unshaded coffee plants grown with fertilizer yield the most coffee, although unfertilized shaded crops generally yield more than unfertilized unshaded crops: the response to fertilizer is much greater in full sun. Although traditional coffee production causes berries to ripen more slowly and produce lower yields, the quality of the coffee is allegedly superior. In addition, the traditional shaded method provides living space for many wildlife species. Proponents of shade cultivation say environmental problems such as deforestation, pesticide pollution, habitat destruction, and soil and water degradation are the side effects of the practices employed in sun cultivation.

The American Birding Association, Smithsonian Migratory Bird Centre, National Arbor Day Foundation, and the Rainforest Alliance have led a campaign for 'shade-grown' and organic coffees, which can be sustainably harvested. Shaded coffee cultivation systems show greater biodiversity than full-sun systems, and those more distant from continuous forest compare rather poorly to undisturbed native forest in terms of habitat value for some bird species.

Another issue concerning coffee is its use of water. According to *New Scientist*, using industrial farming practices, it takes about 140 liters (37 U.S. gal) of water to grow the coffee seeds needed to produce one cup of coffee, and the coffee is often grown in countries where there is a water shortage, such as Ethiopia.

By using sustainable agriculture methods, the amount of water usage can be dramatically reduced, while retaining comparable yields. For comparison, the United States Geological Survey reports that one egg requires an input of 454 liters (120 U.S. gal) of water; one serving of milk requires an input of 246 liters (65 U.S. gal) of water; one serving of rice requires an input of 132 liters (35 U.S. gal) of water; and one glass of wine requires an input of 120 liters (32 U.S. gal) of water.

Coffee grounds may be used for composting or as a mulch. They are especially appreciated by worms and acid-loving plants such as blueberries. Some commercial coffee shops run initiatives to make better use of these grounds, including Starbucks' "Grounds for your Garden" project, and community sponsored initiatives such as "Ground to Ground".

Starbucks sustainability chief Jim Hanna has warned that climate change may significantly impact coffee yields within a few decades. A study by Kew Royal Botanic Gardens concluded that global warming threatens the genetic diversity of Arabica plants found in Ethiopia and surrounding countries.

Production

Figure: *Ripe coffee berries*

Figure: *Unripe coffee berries near Kona, Hawaii*

Processing

Coffee berries and their seeds undergo several processes before they become the familiar roasted coffee. Berries have been traditionally selectively picked by hand; a labour-intensive method, it involves the selection of only the berries at the peak of ripeness. More commonly, crops are strip picked, where all berries are harvested simultaneously regardless of ripeness by person or machine. After picking, green coffee is processed by one of two methods—the *dry process method*, simpler and less labour-intensive as the berries can be strip picked, and the *wet process method*, which incorporates fermentation into the process and yields a mild coffee.

Then they are sorted by ripeness and colour and most often the flesh of the berry is removed, usually by machine, and the seeds are fermented to remove the slimy layer of mucilage still present on the seed. When the fermentation is finished, the seeds are washed with large quantities of fresh water to remove the fermentation residue, which generates massive amounts of coffee wastewater. Finally, the seeds are dried.

The best (but least used) method of drying coffee is using drying tables. In this method, the pulped and fermented coffee is spread thinly on raised beds, which allows the air to pass on all sides of the coffee, and then the coffee is mixed by hand. In this method the drying that takes place is more uniform, and fermentation is less likely. Most African coffee is dried in this manner and certain coffee farms around the world are starting to use this traditional method.

Next, the coffee is sorted, and labelled as green coffee. Another way to let the coffee seeds dry is to let them sit on a concrete patio and rake over them in the sunlight. Some companies use cylinders to pump in heated air to dry the coffee seeds, though this is generally in places where the humidity is very high. Some coffee undergoes a peculiar process, such as kopi luwak. It is made from the seeds of coffee berries which have been eaten by the Asian Palm Civet and other related civets, passing through its digestive tract. This process resulted in coffee seeds with much less bitterness, widely noted as the most expensive coffee in the world with prices reaching $160 per pound.

Roasting

The next step in the process is the roasting of the green coffee. Coffee is usually sold in a roasted state, and with rare exceptions all coffee is roasted before it is consumed. It can be sold roasted by the supplier, or it can be home roasted. The roasting process influences the taste of the beverage by changing the coffee seed both physically and chemically. The seed decreases in weight as moisture is lost and increases in volume, causing it to become less dense. The density of the seed also influences the strength of the coffee and requirements for packaging.

The actual roasting begins when the temperature inside the seed reaches approximately 200 °C (392 °F), though different varieties of seeds differ in moisture and density and therefore roast at different rates. During roasting, caramelization occurs as intense heat breaks down starches, changing them to simple sugars that begin to brown, which alters the colour of the seed. Sucrose is rapidly lost during the roasting process and may disappear entirely in darker roasts. During roasting, aromatic oils and acids weaken, changing the flavour; at 205 °C (401 °F), other oils start to develop. One of these oils, caffeol, is created at about 200 °C (392 °F), which is largely responsible for coffee's aroma and flavour.

Grading the Roasted Seeds

Depending on the colour of the roasted seeds as perceived by the human eye, they will be labelled as light, medium light, medium, medium dark, dark, or very dark. A more accurate method of discerning the degree of roast involves measuring the reflected light from roasted seeds illuminated with a light source in the near infrared spectrum. This elaborate light metre uses a process known as spectroscopy to return a number that consistently indicates the roasted coffee's relative degree of roast or flavour development.

Roast Characteristics

The degree of roast has an effect upon coffee flavour and body. Darker roasts are generally bolder because they have less fibre content and a more sugary flavour. Lighter roasts have a more complex and therefore perceived stronger flavour from aromatic oils and acids otherwise destroyed by longer roasting times. Roasting does not alter the amount of caffeine in the bean, but does give less caffeine when the beans are measured by volume because the beans expand during roasting. A small amount of chaff is produced during roasting from the skin left on the seed after processing. Chaff is usually removed from the seeds by air movement, though a small amount is added to dark roast coffees to soak up oils on the seeds.

Decaffeination

Decaffeination may also be part of the processing that coffee seeds undergo. Seeds are decaffeinated when they are still green. Many methods can remove caffeine from coffee, but all involve either soaking the green seeds in hot water (often called the "Swiss water process") or steaming them, then using a solvent to dissolve caffeine-containing oils. Decaffeination is often done by processing companies, and the extracted caffeine is usually sold to the pharmaceutical industry.

Storage

Coffee is best stored in an airtight container made of ceramic, glass, or non-reactive metal. Higher quality prepackaged coffee usually has a one-way valve which prevents air from entering while allowing the coffee to release gases. Coffee freshness and flavour is preserved when it is stored away from moisture, heat, and light. The ability of coffee to absorb strong smells from food means that it should be kept away from such smells. Storage of coffee in the refrigerator is not recommended due to the presence of moisture which can cause deterioration. Exterior walls of buildings which face the sun may heat the interior of a home, and this heat may damage coffee stored near such a wall. Heat from nearby ovens also harms stored coffee.

In 1931, a method of packing coffee in a sealed vacuum in cans was introduced. The roasted coffee was packed and then 99% of the air was removed, allowing the coffee to be stored indefinitely until the can was opened. Today this method is in mass use for coffee in a large part of the world.

Brewing

Coffee seeds must be ground and brewed to create a beverage. The criteria for choosing a method include flavour and economy. Almost all methods of preparing coffee require that the seeds be ground and then mixed with hot water long enough to allow the flavour to emerge but not so long as to draw out bitter compounds. The liquid can be consumed after the spent grounds are removed. Brewing considerations include the fineness of grind, the way in which the water is to extract the flavour, additional flavourings such as sugar, milk, and spices, and the technique to be used to separate spent grounds. Ideal holding temperatures range from 85–88 °C (185–190 °F) to as high as 93 °C (199 °F) and the ideal serving temperature is 68 to 79 °C (154 to 174 °F).

The roasted coffee seeds may be ground at a roastery, in a grocery store, or in the home. Most coffee is roasted and ground at a roastery and sold in packaged form, though roasted coffee seeds can be ground at home immediately before consumption. It is also possible, though uncommon, to roast raw seeds at home.

The choice of brewing method depends to some extent on the degree to which the coffee seeds have been roasted. Lighter roasted coffee tends to be used for filter coffee as the combination of method and roast style results in higher acidity, complexity, and clearer nuances. Darker roasted coffee is used for espresso because the machine naturally extracts more dissolved solids, causing lighter coffee to become too acidic.

Coffee seeds may be ground in several ways. A burr grinder uses revolving elements to shear the seed; a blade grinder cuts the seeds with blades moving at high speed; and a mortar and pestle crushes the seeds. For most brewing methods a burr grinder is deemed superior because the grind is more even and the grind size can be adjusted.

The type of grind is often named after the brewing method for which it is generally used. Turkish grind is the finest grind, while coffee percolator or French press are the coarsest grinds. The most common grinds are between these two extremes: a medium grind is used in most home coffee-brewing machines.

Coffee may be brewed by several methods. It may be boiled, steeped, or pressurized. Brewing coffee by boiling was the earliest method, and Turkish coffee is an example of this method. It is prepared by grinding or pounding the seeds to a fine powder, then adding it to water and bringing it to the boil for no more than an instant in a pot called a *cezve* or, in Greek, a *br vki*. This produces a strong coffee with a layer of foam on the surface and sediment (which is not meant for drinking) settling at the bottom of the cup.

Coffee percolators and automatic coffeemakers brew coffee using gravity. In an automatic coffeemaker, hot water drips onto coffee grounds that are held in a paper, plastic, or perforated metal coffee filter, allowing the water to seep through the ground coffee while extracting its oils and essences. The liquid drips through the coffee and the filter into a carafe or pot, and the spent grounds are retained in the filter.

In a percolator, boiling water is forced into a chamber above a filter by steam pressure created by boiling. The water then seeps through the grounds, and the process is repeated until terminated by removing from the heat, by an internal timer, or by a thermostat that turns off the heater when the entire pot reaches a certain temperature.

Coffee may be brewed by steeping in a device such as a French press (also known as a *cafetière*, coffee press or coffee plunger). Ground coffee and hot water are combined in a cylindrical vessel and left to brew for a few minutes. A circular filter which fits tightly in the cylinder fixed to a plunger is then pushed down from the top to force the grounds to the bottom. The filter retains the grounds at the bottom as the coffee is poured from the container. Because the coffee grounds are in direct contact with the water, all the coffee oils remain in the liquid, making it a stronger beverage. This method of brewing leaves more sediment than in coffee made by an automatic coffee machine. Supporters of the French press method point out that the sediment issue can be minimized by using the right type of grinder: they claim that a rotary blade grinder cuts the coffee bean into a wide range of sizes, including a fine coffee dust that remains as sludge at the bottom of the cup, while a burr grinder uniformly grinds the beans into consistently-sized grinds, allowing the coffee to settle uniformly and be trapped by the press. 95% of the caffeine is released from the coffee seeds within the first minute of brewing.

The espresso method forces hot pressurized and vaporized water through ground coffee. As a result of brewing under high pressure (ideally between 9–10 atm), the espresso beverage is more concentrated (as much as 10 to 15 times the quantity of coffee to water as gravity-brewing methods can produce) and has a more complex physical and chemical constitution. A well-prepared espresso has a reddish-brown foam called *crema* that floats on the surface. Other pressurized water methods include the moka pot and vacuum coffee maker.

Cold brew coffee is made by steeping coarsely ground seeds in cold water for several hours, then filtering them. This results in a brew lower in acidity than most hot-brewing methods.

Sale and Distribution

Coffee ingestion on average is about a third of that of tap water in North America and Europe. Worldwide, 6.7 million metric tons of coffee were produced annually in 1998–2000, and the forecast is a rise to seven million metric tons annually by 2010.

Brazil remains the largest coffee exporting nation, but Vietnam tripled its exports between 1995 and 1999 and became a major producer of robusta seeds. Indonesia is the third-largest coffee exporter overall and the largest producer of washed arabica coffee. Organic Honduran coffee is a rapidly growing emerging commodity owing to the Honduran climate and rich soil.

In 2013 *The Seattle Times* reported that global coffee prices have dropped more than 50 percent year-over-year. In Thailand Black Ivory coffee beans are fed to elephants, whose digestive enzymes remove much of the beans bitter taste. After being collected from their dung, these beans can sell for as much as $1,100 a kilogram making it the most expensive blend in the world. Another blend, Kopi Luwak is made through a similar process after being digested by the Asian Palm Civet; but it is sold for somewhat less at 100-$600 a pound.

Commodity

Coffee is bought and sold by roasters, investors, and price speculators as a tradable commodity in commodity markets and exchange-traded funds.

Coffee futures contracts for Grade 3 washed arabicas are traded on the New York Mercantile Exchange under ticker symbol KC, with contract deliveries occurring every year in March, May, July, September, and December. Coffee is an example of a product that has been susceptible to significant commodity futures price variations.

Higher and lower grade arabica coffees are sold through other channels. Futures contracts for robusta coffee are traded on the London International Financial Futures and Options Exchange and, since 2007, on the New York Intercontinental Exchange. Coffee has been described by many, including historian Mark Pendergrast, as the world's "second most legally traded commodity." However, this claim has been recently refuted by Pendergrast among others after further research.

Webers Theory of Industrial Location

Alfred Weber, a German economist, enunciated a systematic theory of industrial location in 1909. Weber's theory of location is purely deductive in its approach. He analyzed the factors that determine the location of industry and classified these factors into two divisions. These are:

- Primary causes of regional distribution of industry (regional factors)
- Secondary causes (agglomerative and deglomerative factors) that are responsible for redistribution of industry.
- Primary Causes (Regional Factors)

According to Weber, transport costs and labour costs are the two regional factors on which his pure theory is based. Assuming that there are no other factors that influence the distribution of industry, except transportation costs. Then it is clear that the location of industry will be pulled to those locations which have the lowest transportation costs. The key factors that determine transportation costs are

- *the weight to be transported and*
- *the distance to be covered.*

Weber lists some more factors which influence the transportation costs such as – (a) the type of transportation system and the extent of its use, (b) the nature of the region and kinds of roads, (c) the nature of goods themselves, i.e., the qualities which, besides weight, determine the facility of transportation.

However, the location of the place of production must be determined in relation to the place of consumption and to the most advantageously located material deposits. Thus, 'locational figures' are created. These locational figures depends upon (a) the type of material deposits and (b) the nature of transformation into products.

Weber classifies and calls those raw materials, which are available practically everywhere as 'ubiquities' (like brick-clay, water, etc) and 'localised' (like iron-ore, minerals, wood, etc) which are available only in certain regions. It is clear that localized materials play a more important role on the industry than the ubiquities. Further, regarding the nature of the transformation of materials into products, Weber categorized the raw materials as 'pure' and 'weight losing'. Pure materials impart their total weight to the products (eg. cotton, wool, etc) and the materials are said to be 'weight losing' if only a part enters into the product (eg. wood, coal, etc.). Hence, the location of industries using weight-losing materials is drawn towards their deposits and that of industries using pure-materials towards the consumption centres.

Weber further examines the cause of deviation of industrial location from the centres of least transport costs. The existence of differences in labour costs leads an industry to deviate from the optimal point of transport orientation. Geographical distribution of the population would give rise to differences in wages for labour. Naturally, the transport oriented location of an industry is drawn out and attracted towards the cheaper labour centres. Such migration of an industry from a point of minimum transport costs to a cheaper labour centre may be likely to occur only where the savings in the cost of labour are larger than the additional costs of transport which it ought to incur.

- Secondary Causes (Agglomerative and Deglomerative Factors)

An agglomerative factor is an advantage or a cheapening of production or marketing which results from the fact that production is carried on at one place. A deglomerative factor is a cheapening of production which results from the decentralization of production i.e., production in more than one place. To some extent these agglomerative and deglomerative factors also contribute to local accumulation and distribution of industry. These factors will operate only within the general framework formed by the two regional factors, i.e., costs of transportation and costs of labour. The advantages which could be derived in this context are external economies.

The pulls which the agglomerative factors possess to attract an industry to a particular point are mainly dependent on two factors. Firstly, on 'the index of manufacture' (the proportion of manufacturing costs to the total weight of the product) and secondly, on the 'locational weight' (the total weight to be transported during all the stages of production). To deduce a general principle, Weber uses the concept of "co-efficient of manufacture" which is the ratio of manufacturing cost to locational weight. Agglomeration is encouraged with high co-efficient of manufacture and deglomeration with low co-efficient of manufacture and these tendencies are inherent in their nature.

Split Location

Productive activities could be divided depending on the nature of raw-materials, industry and market. Weber considers the location for an industry at more than one place. According to Weber, a split of production into several locations will be the rule for productive process which can technically be split. For instance, the first stage of production may be near the raw material deposits and the subsequent stages near the place of final consumption. Likewise, in a paper industry the manufacture of pulp may be carried on near the supplies of the raw materials and the second stage of paper manufacture near the consumption outlet.

Locational Coupling

Weber also conceived the advantages of setting up different types of industries in the same locality. The production of quite different articles may be combined in one plant because several raw materials may diverge from a common source. This may be either due to technical or economic reasons: for instance, certain chemical industries, garments factories which manufacture over-coats, shawls, blouses, etc. Locational coupling may also occur due to connection through materials. If the by-product of an industry happens to be the raw material of another industry, then the two industries may select a single place of location. For instance, the dye-stuff industry is connected with other industries using coke, because coal tar (upon which the dye-stuff industry is based) is a by-product of the burning coke.

Criticisms

Weber's theory of location has been criticized on various grounds which may be summarized as follows:

1. Weber has been criticized for his unrealistic approach and deductive reasoning. According to Sargant Florence; vague generalizations cannot provide suitable solutions to the theory of location as non-economic considerations will also influence which are not mentioned in the pure theory. He says that Weber's theory fails to explain locations resulting from historical and social forces.
2. A.Predohl criticizes Weber's theory as more a selective theory than a deductive theory. The very distinction between primary and secondary is itself artificial, illogical and arbitrary.
3. Weber assumes fixed labour centres and unlimited supplies of labour which are unrealistic. The rise of industry may create new labour centres and we cannot assume unlimited labour supplies at any centre.
4. In a competitive market structure, the assumption of fixed points of consumption is unrealistic. Country-wise scattering, usually, of consuming public is a reality and there may be a shift in the consuming centres with a shift in industrial population.
5. A. Robinson also considers Weber's division of raw materials into 'ubiquities' and 'localised' as artificial.

Weber's deductive theory of location, in spite of the shortcomings, is the only theory which has been enjoying the universal acceptance and application, as all the other alternative suggestions are neither complete nor comprehensive.

Chapter 7

Major Minerals

Iron Ore

Iron ores are rocks and minerals from which metallic iron can be economically extracted. The ores are usually rich in iron oxides and vary in colour from dark grey, bright yellow, deep purple, to rusty red. The iron itself is usually found in the form of magnetite (Fe_3O_4, 72.4% Fe), hematite (Fe2O3, 69.9% Fe), goethite (FeO(OH), 62.9% Fe), limonite ($FeO(OH).n(H_2O)$) or siderite ($FeCO_3$, 48.2% Fe).

Ores carrying very high quantities of hematite or magnetite (greater than ~60% iron) are known as "natural ore" or "direct shipping ore", meaning they can be fed directly into iron-making blast furnaces. Iron ore is the raw material used to make pig iron, which is one of the main raw materials to make steel. 98% of the mined iron ore is used to make steel. Indeed, it has been argued that iron ore is "more integral to the global economy than any other commodity, except perhaps oil".

Sources

Metallic iron is virtually unknown on the surface of the Earth except as iron-nickel alloys from meteorites and very rare forms of deep mantle xenoliths. Although iron is the fourth most abundant element in the Earth's crust, comprising about 5%, the vast majority is bound in silicate or more rarely carbonate minerals. The thermodynamic barriers to separating pure iron from these minerals are formidable and energy intensive, therefore all sources of iron used by human industry exploit comparatively rarer iron oxide minerals, primarily hematite.

Prior to the industrial revolution, most iron was obtained from widely available goethite or bog ore, for example during the American

Revolution and the Napoleonic wars. Prehistoric societies used laterite as a source of iron ore. Historically, much of the iron ore utilized by industrialized societies has been mined from predominantly hematite deposits with grades of around 70% Fe. These deposits are commonly referred to as "direct shipping ores" or "natural ores". Increasing iron ore demand, coupled with the depletion of high-grade hematite ores in the United States, after World War II led to development of lower-grade iron ore sources, principally the utilization of magnetite and taconite. (Taconite is a rock whose iron content, commonly present as finely dispersed magnetite, is generally 25 to 30%.)

Iron ore mining methods vary by the type of ore being mined. There are four main types of iron ore deposits worked currently, depending on the mineralogy and geology of the ore deposits. These are magnetite, titanomagnetite, massive hematite and pisolitic ironstone deposits.

Production and Consumption

Iron is the world's most commonly used metal - steel, of which iron ore is the key ingredient, representing almost 95% of all metal used per year. It is used primarily in structural engineering applications and in maritime purposes, automobiles, and general industrial applications (machinery). Iron-rich rocks are common worldwide, but ore-grade commercial mining operations are dominated by the countries listed in the table aside. The major constraint to economics for iron ore deposits is not necessarily the grade or size of the deposits, because it is not particularly hard to geologically prove enough tonnage of the rocks exist. The main constraint is the position of the iron ore relative to market, the cost of rail infrastructure to get it to market and the energy cost required to do so.

Mining iron ore is a high volume low margin business, as the value of iron is significantly lower than base metals. It is highly capital intensive, and requires significant investment in infrastructure such as rail in order to transport the ore from the mine to a freight ship. For these reasons, iron ore production is concentrated in the hands of a few major players. World production averages two billion metric tons of raw ore annually. The world's largest producer of iron ore is the Brazilian mining corporation Vale, followed by Anglo-Australian companies BHP Billiton and Rio Tinto Group. A further Australian supplier, Fortescue Metals Group Ltd has helped bring Australia's production to second in the world.

The seaborne trade in iron ore, that is, iron ore to be shipped to other countries, was 849m tonnes in 2004. Australia and Brazil dominate the seaborne trade, with 72% of the market. BHP, Rio and Vale control 66% of this market between them. In Australia iron ore is won from three main sources: pisolite "channel iron deposit" ore derived by mechanical erosion of primary banded-iron formations and accumulated in alluvial channels such as at Pannawonica, Western Australia; and the dominant metasomatically-altered banded iron formation related ores such as at Newman, the Chichester Range, the Hamersley Range and Koolyanobbing, Western Australia. Other types of ore are coming to the fore recently, such as oxidised ferruginous hardcaps, for instance laterite iron ore deposits near Lake Argyle in Western Australia.

The total recoverable reserves of iron ore in India are about 9,602 million tones of hematite and 3,408 million tones of magnetite. Chhattisgarh, Madhya Pradesh, Karnataka, Jharkhand, Odisha, Goa, Maharashtra, Andhra Pradesh, Kerala, Rajasthan and Tamil Nadu are the principal Indian producers of iron ore. World consumption of iron ore grows 10% per annum on average with the main consumers being China, Japan, Korea, the United States and the European Union. China is currently the largest consumer of iron ore, which translates to be the world's largest steel producing country. It is also the largest importer, buying 52% of the seaborne trade in iron ore in 2004. China is followed by Japan and Korea, which consume a significant amount of raw iron ore and metallurgical coal. In 2006, China produced 588 million tons of iron ore, with an annual growth of 38%.

Iron Ore Market

Over the last 40 years, iron ore prices have been decided in closed-door negotiations between the small handful of miners and steelmakers which dominate both spot and contract markets. Traditionally, the first deal reached between these two groups sets a *benchmark* to be followed by the rest of the industry.

In recent years, however, this benchmark system has begun to break down, with participants along both demand and supply chains calling for a shift to short term pricing. Given that most other commodities already have a mature market-based pricing system, it is natural for iron ore to follow suit. To answer increasing market demands for more transparent pricing, a number of financial exchanges and/or clearing houses around the world have offered iron ore swaps clearing. The CME group, SGX (Singapore Exchange), London Clearing House (LCH.Clearnet), NOS Group and ICEX (Indian Commodities Exchange) all offer cleared swaps based on The Steel Index's (TSI) iron ore transaction data. The CME also offers a Platts based swap, in addition to their TSI swap clearing. The ICE (Intercontinental Exchange) offers a Platts based swap clearing service also. The swaps market has grown quickly, with liquidity clustering around TSI's pricing. By April 2011, over US$5.5 billion worth of iron ore swaps have been cleared basis TSI prices. By August 2012, in excess of one million tonnes of swaps trading per day was taking place regularly, basis TSI.

A relatively new development has also been the introduction of iron ore options, in addition to swaps. The CME group has been the venue most utilised for clearing of options written against TSI, with open interest at over 12,000 lots in August 2012.

Singapore Mercantile Exchange (SMX) has launched the world first global iron ore futures contract, based on the Metal Bulletin Iron Ore Index (MBIOI) which utilizes daily price data from a broad spectrum of industry participants and independent Chinese steel consultancy and data provider Shanghai Steelhome's widespread contact base of steel producers and iron ore traders across China. The futures contract has seen monthly volumes over 1.5 million tonnes after eight months of trading. This move follows a switch to index-based quarterly pricing by the world's three largest iron ore miners - Vale, Rio Tinto and BHP Billiton - in early 2010, breaking a 40-year tradition of benchmark annual pricing.

Available Iron ore Resources

Available World Iron ore Resources

The Iron ore reserves at present seem quite vast, but some are starting to suggest that the math of continual exponential increase in consumption can even make this resource seem quite finite. For instance, Lester Brown of the Worldwatch Institute has suggested iron ore could run out within 64 years based on an extremely conservative extrapolation of 2% growth per year.

Australia

Geoscience Australia calculates that the country's "economic demonstrated resources" of iron currently amount to 24 gigatonnes, or 24 billion tonnes. The current production rate from the Pilbara region of Western Australia is approximately 430 million tonnes a year and rising. Experts Dr Gavin Mudd (Monash University) and Jonathon Law (CSIRO) expect it to be gone within 30 to 50 years (Mudd) and 56 years (Law). These estimates require on-going review to take into account shifting demand for lower grade iron ore and improving mining and recovery techniques.

Pilbara Deposit

In 2011, leading Pilbara based iron ore miners - Rio Tinto, BHP Billiton and Fortescue Metals Group - all announced significant capital investment in the development of existing and new mines and associated infrastructure (rail and port). Collectively this would amount to the production of 1,000 million tonnes per year (Mt/y) by 2020. Practically that would require a doubling of production capacity from a current production level of 470 Mt/y to 1,000 Mt/y (an increase of 530 Mt/y). These figures are based on the current production rates of Rio 220 Mt/y, BHP 180 Mt/y, FMG 55 Mt/y and Other 15 Mt/y increasing to Rio 353 Mt/y, BHP 356 Mt/y, FMG 155 Mt/y and Other 140 Mt/y (the latter 140 Mt/y is based on planned production from recent industry entrants Hancock, Atlas and Brockman through Port Hedland and API and others through the proposed Port of Anketell).

A production rate of 1,000 Mt/y would require a significant increase in production from existing mines and the opening of a significant number of new mines. Further, a significant increase in the capacity of rail and port infrastructure would also be required. For example, Rio would be required to expand its port operations at Dampier and Cape Lambert by 140 Mt/y (from 220 Mt/y to 360 Mt/y). BHP would be required to expand its Port Hedland port operations by 180 Mt/y (from 180 Mt/y to 360 Mt/y). FMG would be required to expand its port operations at Port Hedland by 100 Mt/y (from 55 Mt/y to 155 Mt/y). That's an increase of 420 Mt/y in port capacity by the three majors Rio, BHP and FMG and about at least 110 Mt/y from the non-major producers. Based on the rule-of-thumb of 50 Mt/y per car dumper, reclaimer and ship-loader the new production would require approximately 10 new car dumpers, reclaimers and ship-loaders.

New rail capacity would also be required. Based on the rule-of-thumb of 100 Mt/y per rail line, increasing production by approximately 500 Mt/y would require 5 new single rail lines. One scenario is an extra rail line for all the majors: BHP (from double to triple track), Rio (double to triple track), FMG (single to double track) and at least two new lines. New lines have been proposed by Hancock to service the Roy Hill mine and QR National to service non-major producers.

A 1,000 Mt/y production rate needs to be further considered by proponents and government. Areas of further consideration include new port space at Anketell to service the West Pilbara mines, growth at Port Hedland (BHP has announced the development of an outer harbour at Port Hedland), rail rationalisation and the regulatory approval requirements for opening and maintaining a ground disturbance footprint that supports 1,000 Mt/y of production including, amongst other things, native title, aboriginal heritage and environmental protection outcomes.

Manganese

Manganese is a chemical element with symbol Mn and atomic number 25. It is not found as a free element in nature; it is often found in combination with iron, and in many minerals. Manganese is a metal with important industrial metal alloy uses, particularly in stainless steels.

Historically, manganese is named for various black minerals (such as pyrolusite) from the same region of Magnesia in Greece which gave names to similar-sounding magnesium, Mg, and magnetite, an ore of the element iron, Fe. By the mid-18th century, Swedish chemist Carl Wilhelm Scheele had used pyrolusite to produce chlorine. Scheele and others were aware that pyrolusite (now known to be manganese dioxide) contained a new element, but they were not able to isolate it. Johan Gottlieb Gahn was the first to isolate an impure sample of manganese metal in 1774, by reducing the dioxide with carbon.

Manganese phosphating is used as a treatment for rust and corrosion prevention on steel. Depending on their oxidation state, manganese ions have various colours and are used industrially as pigments. The permanganates of alkali and alkaline earth metals are powerful oxidizers. Manganese dioxide is used as the cathode (electron acceptor) material in zinc-carbon and alkaline batteries.

In biology, manganese(II) ions function as cofactors for a large variety of enzymes with many functions. Manganese enzymes are particularly essential in detoxification of superoxide free radicals in organisms that must deal with elemental oxygen. Manganese also functions in the oxygen-evolving complex of photosynthetic plants. The element is a required trace mineral for all known living organisms. In larger amounts, and apparently with far greater activity by inhalation, it can cause a poisoning syndrome in mammals, with neurological damage which is sometimes irreversible.

Occurrence and Production

Manganese makes up about 1000 ppm (0.1%) of the Earth's crust, making it the 12th most abundant element there. Soil contains 7–9000 ppm of manganese with an average of 440 ppm. Seawater has only 10 ppm manganese and the atmosphere contains 0.01 $\mu g/m^3$. Manganese occurs principally as pyrolusite (MnO_2), braunite,

$(Mn^{2+}Mn^{3+}{}_6)(SiO_{12})$, psilomelane $(Ba,H_2O)_2Mn_5O_{10}$, and to a lesser extent as rhodochrosite ($MnCO_3$).

The most important manganese ore is pyrolusite (MnO_2). Other economically important manganese ores usually show a close spatial relation to the iron ores. Land-based resources are large but irregularly distributed. About 80% of the known world manganese resources are found in South Africa; other important manganese deposits are in Ukraine, Australia, India, China, Gabon and Brazil. In 1978, 500 billion tons of manganese nodules were estimated to exist on the ocean floor. Attempts to find economically viable methods of harvesting manganese nodules were abandoned in the 1970s.

In South Africa most identified deposits are located near Hotazel in the Northern Cape Province with an estimated 15 billion tons in 2011. In 2011 South Africa was the world's largest producer of manganese producing 3.4 million tons.

Manganese is mined in South Africa, Australia, China, Brazil, Gabon, Ukraine, India and Ghana and Kazakhstan. US Import Sources (1998–2001): Manganese ore: Gabon, 70%; South Africa, 10%; Australia, 9%; Mexico, 5%; and other, 6%. Ferromanganese: South Africa, 47%; France, 22%; Mexico, 8%; Australia, 8%; and other, 15%. Manganese contained in all manganese imports: South Africa, 31%; Gabon, 21%; Australia, 13%; Mexico, 8%; and other, 27%.

For the production of ferromanganese, the manganese ore is mixed with iron ore and carbon, and then reduced either in a blast furnace or in an electric arc furnace. The resulting ferromanganese has a manganese content of 30 to 80%. Pure manganese used for the production of iron-free alloys is produced by leaching manganese ore with sulfuric acid and a subsequent electrowinning process.

A more progressive extraction process involves directly reducing manganese ore in a heap leach. This is done by percolating natural gas through the bottom of the heap; the natural gas provides the heat (needs to be at least 850 °C) and the reducing agent (carbon monoxide). This reduces all of the manganese ore to manganese oxide (MnO), which is a leachable form. The ore then travels through a grinding circuit to reduce the particle size of the ore to between 150–250 μm, this increases the surface area to aid in the leaching process. The ore is then added to a leach tank, which contains sulfuric acid and ferrous iron (Fe^{2+}) in a 1.6:1 ratio. The iron reacts with the manganese dioxide to form iron hydroxide and elemental manganese. This process yields approximately 92% recovery of the manganese. For further purification, the manganese can then be sent to an electrowinning facility.

In 1972 the CIA's Project Azorian, through billionaire Howard Hughes, commissioned the ship *Hughes Glomar Explorer* with the cover story of harvesting manganese nodules from the sea floor. That triggered a rush of activity to attempt to collect manganese nodules, which was not actually practical. The real mission of *Hughes Glomar Explorer* was to raise a sunken Soviet submarine, the K-129, with the goal of retrieving Soviet code books.

Bauxite

Bauxite, an aluminium ore, is the world's main source of aluminium. It consists mostly of the minerals gibbsite $Al(OH)_3$, boehmite γ-AlO(OH) and diaspore α-AlO(OH), mixed with the two iron oxides goethite and haematite, the clay mineral kaolinite and small amounts of anatase TiO_2. Bauxite was named by the French geologist Pierre Berthier in 1821 after the village of Les Baux in Provence, southern France, where he discovered it and was the first to recognize that it contained aluminium.

Production Trends

In 2009, Australia was the top producer of bauxite with almost one-third of the world's production, followed by China, Brazil, India, and Guinea. Although aluminium demand is rapidly increasing, known reserves of its bauxite ore are sufficient to meet the worldwide demands for aluminium for many centuries. Increased aluminium recycling, which has the advantage of lowering the cost in electric power in producing aluminium, will considerably extend the world's bauxite reserves.

Processing

Bauxite is usually strip mined because it is almost always found near the surface of the terrain, with little or no overburden. Approximately 70% to 80% of the world's dry bauxite production is processed first into alumina, and then into aluminium by electrolysis as of 2010. Bauxite rocks are typically classified according to their intended commercial application: metallurgical, abrasive, cement, chemical, and refractory.

Usually, bauxite ore is heated in a pressure vessel along with a sodium hydroxide solution at a temperature of 150 to 200 °C. At these temperatures, the aluminium is dissolved as an aluminate (the Bayer process). After separation of ferruginous residue (red mud) by filtering, pure gibbsite is precipitated when the liquid is cooled, and then seeded with fine-grained aluminium hydroxide. The gibbsite is usually converted into aluminium oxide, Al_2O_3, by heating. This mineral is dissolved at a temperature of about 960 °C in molten cryolite. Next, this molten substance can yield metallic aluminium by passing an electric current through it in the process of electrolysis, which is called the Hall–Héroult process, named after its American and French discoverers.

Prior to the invention of this process in 1886, elemental aluminium was made by heating ore along with elemental sodium or potassium in a vacuum. The method was complicated and consumed materials that were themselves expensive at that time. This made early elemental aluminium more expensive than gold.

Copper

Copper is a chemical element with symbol Cu (from Latin: *cuprum*) and atomic number 29. It is a ductile metal with very high thermal and electrical conductivity. Pure copper is soft and malleable; a freshly exposed surface has a reddish-orange colour. It is used as a conductor of heat and electricity, a building material, and a constituent of various metal alloys.

The metal and its alloys have been used for thousands of years. In the Roman era, copper was principally mined on Cyprus, hence the origin of the name of the metal as *ñyprium* (metal of Cyprus), later shortened to *ñuprum*. Its compounds are commonly encountered as copper(II) salts, which often impart blue or green colours to minerals such as azurite and turquoise and have been widely used historically as pigments. Architectural structures built with copper corrode to give green verdigris (or patina). Decorative art prominently features copper, both by itself and as part of pigments.

Copper is essential to all living organisms as a trace dietary mineral because it is a key constituent of the respiratory enzyme complex cytochrome c oxidase. In molluscs and crustacea copper is a constituent of the blood pigment hemocyanin, which is replaced by the iron-complexed hemoglobin in fish and other vertebrates. The main areas where copper is found in humans are liver, muscle and bone. Copper compounds are used as bacteriostatic substances, fungicides, and wood preservatives.

Production

Most copper is mined or extracted as copper sulfides from large open pit mines in porphyry copper deposits that contain 0.4 to 1.0% copper. Examples include Chuquicamata in Chile, Bingham Canyon

Mine in Utah, United States and El Chino Mine in New Mexico, United States. According to the British Geological Survey, in 2005, Chile was the top mine producer of copper with at least one-third world share followed by the United States, Indonesia and Peru. Copper can also be recovered through the In-situ leach process. Several sites in the state of Arizona are considered prime candidates for this method. The amount of copper in use is increasing and the quantity available is barely sufficient to allow all countries to reach developed world levels of usage.

Reserves

Copper has been in use at least 10,000 years, but more than 96% of all copper ever mined and smelted has been extracted since 1900, and more than half was extracted in only the last 24 years. As with many natural resources, the total amount of copper on Earth is vast (around 10^{14} tons just in the top kilometre of Earth's crust, or about 5 million years' worth at the current rate of extraction). However, only a tiny fraction of these reserves is economically viable, given present-day prices and technologies. Various estimates of existing copper reserves available for mining vary from 25 years to 60 years, depending on core assumptions such as the growth rate. Recycling is a major source of copper in the modern world. Because of these and other factors, the future of copper production and supply is the subject of much debate, including the concept of peak copper, analogous to peak oil.

The price of copper has historically been unstable, and it sextupled from the 60-year low of US$0.60/lb (US$1.32/kg) in June 1999 to US$3.75 per pound (US$8.27/kg) in May 2006. It dropped to US$2.40/lb (US$5.29/kg) in February 2007, then rebounded to US$3.50/lb (US$7.71/kg) in April 2007. In February 2009, weakening global demand and a steep fall in commodity prices since the previous year's highs left copper prices at US$1.51/lb.

Methods

The concentration of copper in ores averages only 0.6%, and most commercial ores are sulfides, especially chalcopyrite ($CuFeS_2$) and to a lesser extent chalcocite (Cu_2S). These minerals are concentrated from crushed ores to the level of 10–15% copper by froth flotation or bioleaching. Heating this material with silica in flash smelting removes much of the iron as slag. The process exploits the greater ease of converting iron sulfides into its oxides, which in turn react with the silica to form the silicate slag, which floats on top of the heated mass. The resulting *copper matte* consisting of Cu_2S is then roasted to convert all sulfides into oxides:

$$2\ Cu_2S + 3\ O_2 \rightarrow 2\ Cu_2O + 2\ SO_2$$

The cuprous oxide is converted to *blister* copper upon heating:

$$2\ Cu_2O \rightarrow 4\ Cu + O_2$$

The Sudbury matte process converted only half the sulfide to oxide and then used this oxide to remove the rest of the sulfur as oxide. It was then electrolytically refined and the anode mud exploited for the platinum and gold it contained. This step exploits the relatively easy reduction of copper oxides to copper metal. Natural gas is blown across the blister to remove most of the remaining oxygen and electrorefining is performed on the resulting material to produce pure copper:

$$Cu^{2+} + 2\ e^- \rightarrow Cu$$

Recycling

Like aluminium, copper is 100% recyclable without any loss of quality, regardless of whether it is in a raw state or contained in a manufactured product. In volume, copper is the third most recycled metal after iron and aluminium. It is estimated that 80% of the copper ever mined is still in use today. According to the International Resource Panel's Metal Stocks in Society report, the global per capita stock of copper in use in society is 35–55 kg. Much of this is in more-developed countries (140–300 kg per capita) rather than less-developed countries (30–40 kg per capita).

The process of recycling copper is roughly the same as is used to extract copper but requires fewer steps. High purity scrap copper is melted in a furnace and then reduced and cast into billets and ingots; lower purity scrap is refined by electroplating in a bath of sulfuric acid.

Alloys

Numerous copper alloys exist, many with important uses. Brass is an alloy of copper and zinc. Bronze usually refers to copper-tin alloys, but can refer to any alloy of copper such as aluminium bronze. Copper is one of the most important constituents of carat silver and gold alloys, and carat solders are used in the jewellery industry, modifying the colour, hardness and melting point of the resulting alloys.

The alloy of copper and nickel, called cupronickel, is used in low-denomination coins, often for the outer cladding. The US 5-cent coin called a *nickel* consists of 75% copper and 25% nickel and has a homogeneous composition. The alloy consisting of 90% copper and 10% nickel is remarkable for its resistance to corrosion and is used in various parts that are exposed to seawater.

Alloys of copper with aluminium (about 7%) have a pleasant golden colour and are used in decorations. Some lead-free solders consist of tin alloyed with a small proportion of copper and other metals.

Mica

The mica group of sheet silicate (phyllosilicate) minerals includes several closely related materials having close to perfect basal cleavage. All are monoclinic, with a tendency towards pseudohexagonal crystals, and are similar in chemical composition. The nearly perfect cleavage, which is the most prominent characteristic of mica, is explained by the hexagonal sheet-like arrangement of its atoms.

The word *mica* is derived from the Latin word, *mica,* meaning *a crumb*, and probably influenced by *micare*, to glitter.

Classification

Chemically, micas can be given the general formula

$$X_2Y_{4-6}Z_8O_{20}(OH,F)_4$$

in which

X is K, Na, or Ca or less commonly Ba, Rb, or Cs;

Y is Al, Mg, or Fe or less commonly Mn, Cr, Ti, Li, etc.;

Z is chiefly Si or Al, but also may include Fe^{3+} or Ti.

Structurally, micas can be classed as dioctahedral ($Y = 4$) and trioctahedral ($Y = 6$). If the X ion is K or Na, the mica is a *common* mica, whereas if the X ion is Ca, the mica is classed as a *brittle* mica.

Trioctahedral Micas

Common micas:

- Biotite
- Lepidolite
- Muscovite
- Phlogopite
- Zinnwaldite

Brittle micas:

- Clintonite

Interlayer Deficient Micas

Very fine-grained micas, which typically show more variation in ion and water content, are informally termed "clay micas". They include

- Hydro-muscovite with H_3O^+ along with K in the *X* site;
- Illite with a K deficiency in the *X* site and correspondingly more Si in the *Z* site;
- Phengite with Mg or Fe^{2+} substituting for Al in the *Y* site and a corresponding increase in Si in the *Z* site.

Occurrence and Production

Mica is widely distributed and occurs in igneous, metamorphic and sedimentary regimes. Large crystals of mica used for various applications are typically mined from granitic pegmatites.

Until the 19th century, large crystals of mica were quite rare and expensive as a result of the limited supply in Europe. However, their price dramatically dropped when large reserves were found and mined in Africa and South America during the early 19th century. The largest documented single crystal of mica (phlogopite) was found in Lacey mine, Ontario, Canada; it measured 10×4.3×4.3 m and weighed about 330 tonnes. Similar-sized crystals were also found in Karelia, Russia.

The British Geological Survey reported that as of 2005, Koderma district in Jharkhand state in India had the largest deposits of mica in the world. China was the top producer of mica with almost a third of the global share, closely followed by the US, South Korea and Canada. Large deposits of sheet mica were mined in New England from the 19th century to the 1970s. Large mines existed in Connecticut, New Hampshire, and Maine. Scrap and flake mica is produced all over the world. In 2010, the major producers were Russia (100,000 tonnes), Finland (68,000 t), United States (53,000 t), South Korea (50,000 t), France (20,000 t) and Canada (15,000 t). The total production was 350,000 t, although no reliable data were available for China. Most sheet mica was produced in India (3,500 t) and Russia (1,500 t). Flake mica comes from several sources: the metamorphic rock called schist as a byproduct of processing feldspar and kaolin resources, from placer deposits, and from pegmatites. Sheet mica is considerably less abundant than flake and scrap mica, and is occasionally recovered from mining scrap and flake mica. The most important sources of sheet mica are pegmatite deposits. Sheet mica prices vary with grade and can range from less than $1 per kilogram for low-quality mica to more than $2,000 per kilogram for the highest quality.

Properties and Uses

The mica group represents 37 phyllosilicate minerals that have a layered or platy texture. The commercially important micas are muscovite and phlogopite, which are used in a variety of applications. Mica's value is based on several of its unique physical properties. The crystalline structure of mica forms layers that can be split or delaminated into thin sheets usually causing foliation in rocks.

These sheets are chemically inert, dielectric, elastic, flexible, hydrophilic, insulating, lightweight, platy, reflective, refractive, resilient, and range in opacity from transparent to opaque. Mica is stable when exposed to electricity, light, moisture, and extreme temperatures. It has superior electrical properties as an insulator and as a dielectric, and can support an electrostatic field while dissipating minimal energy in the form of heat; it can be split very thin (0.025 to 0.125 millimetres or thinner) while maintaining its electrical properties, has a high dielectric breakdown, is thermally stable to 500 °C, and is resistant to corona discharge.

Muscovite, the principal mica used by the electrical industry, is used in capacitors that are ideal for high frequency and radio frequency. Phlogopite mica remains stable at higher temperatures (to 900 °C) and is used in applications in which a combination of high-heat stability and electrical properties is required. Muscovite and phlogopite are used in sheet and ground forms.

Gold

Gold is a chemical element with symbol Au and atomic number 79. It is a bright yellow dense, soft, malleable and ductile metal. The properties remain when exposed to air or water. Chemically, gold is a transition metal and a group 11 element. It is one of the least reactive chemical elements, and is solid under standard conditions. The metal therefore occurs often in free elemental (native) form, as nuggets or grains, in rocks, in veins and in alluvial deposits. It occurs in a solid solution series with the native element silver (as electrum) and also naturally alloyed with copper and palladium. Less commonly, it occurs in minerals as gold compounds, often with tellurium (gold tellurides).

Gold's atomic number of 79 makes it one of the higher atomic number elements that occur naturally in the universe, and is traditionally thought to have been produced in supernova nucleosynthesis to seed the dust from which the Solar System formed. Because the Earth was molten when it was just formed, almost all of the gold present in the Earth sank into the planetary core. Therefore most of the gold that is present today in the Earth's crust and mantle is thought to have been delivered to Earth later, by asteroid impacts during the late heavy bombardment, about 4 billion years ago.

Gold resists attacks by individual acids, but it can be dissolved by aqua regia (nitro-hydrochloric acid), so named because it dissolves gold into a soluble gold tetrachloride cation. Gold compounds also dissolve in alkaline solutions of cyanide, which have been used in mining. It dissolves in mercury, forming amalgam alloys; it is insoluble in nitric acid, which dissolves silver and base metals, a property that has long been used to confirm the presence of gold in items, giving rise to the term *acid test*. This metal has been a valuable and highly sought-after precious metal for coinage, jewellery, and other arts since long before the beginning of recorded history. In the past, a gold standard was often implemented as a monetary policy within and between nations, but gold coins ceased to be minted as a circulating currency in the 1930s, and the world gold standard was finally abandoned for a fiat currency system after 1976. The historical value of gold was rooted in its medium rarity, easy handling and minting, easy smelting, non-corrosiveness, distinct colour, and non-reactivity to other elements.

A total of 174,100 tonnes of gold have been mined in human history, according to GFMS as of 2012. This is roughly equivalent to 5.6 billion troy ounces or, in terms of volume, about 9200 m^3, or a cube 21 m on a side. The world consumption of new gold produced is about 50% in jewellery, 40% in investments, and 10% in industry. Gold's high malleability, ductility, resistance to corrosion and most other chemical reactions, and conductivity of electricity have led to its continued use in corrosion resistant electrical connectors in all types of computerized devices (its chief industrial use). Gold is also used in infrared shielding, coloured-glass production, and gold leafing. Certain gold salts are still used as anti-inflammatories in medicine.

Occurrence

Gold's atomic number of 79 makes it one of the higher atomic number elements that occur naturally. Although traditionally, gold is thought to have formed by supernova nucleosynthesis., a new theory suggests that gold and other elements heavier than iron are made by the collision of neutron stars instead. Either way, satellite spectrometres in theory detect the resulting gold, "but we have no spectroscopic evidence that [such] elements have truly been produced." These gold nucleogenesis theories hold that the resulting explosions scattered metal-containing dusts (including heavy elements like gold) into the region of space in which they later condensed into our solar system and the Earth. Because the Earth was molten when it was just formed, almost all of the gold present on Earth sank into the core. Most of the gold that is present today in the Earth's crust and mantle is thought to have been delivered to Earth later, by asteroid impacts during the late heavy bombardment.

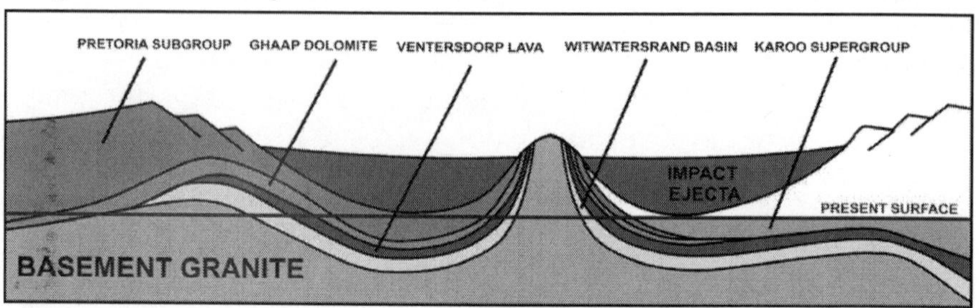

Figure: *A schematic diagram of a NE (left) to SW (right) cross-section through the 2.020 billion year old Vredefort impact crater in South Africa and how it distorted the contemporary geological structures. The present erosion level is shown. Johannesburg is located where the Witwatersrand Basin (the yellow layer) is exposed at the "present surface" line, just inside the crater rim, on the left. Not to scale.*

The asteroid that formed Vredefort crater 2.020 billion years ago is often credited with seeding the Witwatersrand basin in South Africa with the richest gold deposits on earth. However, the gold bearing Witwatersrand rocks were laid down between 700 and 950 million years before the Vredefort impact. These gold bearing rocks had furthermore been covered by a thick layer of Ventersdorp lavas, and the Transvaal Supergroup of rocks before the meteor struck. What the Vredefort impact achieved, however, was to distort the Witwatersrand basin in such a way that the gold bearing rocks were brought to the present erosion surface in Johannesburg, on the Witwatersrand, just inside the rim of the original 300 km diameter crater caused by the meteor strike. The discovery of the deposit in 1886 launched the Witwatersrand Gold Rush. Nearly 50% of all the gold ever mined on earth has been extracted from these Witwatersrand rocks.

On Earth, gold is found in ores in rock formed from the Precambrian time onward. It most often occurs as a native metal, typically in a metal solid solution with silver (i.e. as a gold silver alloy). Such alloys usually have a silver content of 8–10%. Electrum is elemental gold with more than 20% silver. Electrum's colour runs from golden-silvery to silvery, dependent upon the silver content. The more silver, the lower the specific gravity. Native gold occurs as very small to microscopic particles embedded in rock, often together with quartz or sulfide minerals such as "Fool's Gold", which is a pyrite. These are called lode deposits. The metal in a native state is also found in the form of free flakes, grains or larger nuggets that have been eroded from rocks and end up in alluvial deposits called placer deposits.

Such free gold is always richer at the surface of gold-bearing veins owing to the oxidation of accompanying minerals followed by weathering, and washing of the dust into streams and rivers, where it collects and can be welded by water action to form nuggets.

Figure: *Relative sizes of an 860 kg block of gold ore, and the 30 g of gold that can be extracted from it. Toi gold mine, Japan.*

Figure: *Gold left behind after a pyrite cube was oxidized to hematite. Note cubic shape of cavity.*

Gold sometimes occurs combined with tellurium as the minerals calaverite, krennerite, nagyagite, petzite and sylvanite, and as the rare bismuthide maldonite (Au_2Bi) and antimonide aurostibite ($AuSb_2$). Gold also occurs in rare alloys with copper, lead, and mercury: the minerals auricupride (Cu_3Au), novodneprite ($AuPb_3$) and weishanite ($(Au, Ag)_3Hg_2$).

Recent research suggests that microbes can sometimes play an important role in forming gold deposits, transporting and precipitating gold to form grains and nuggets that collect in alluvial deposits.

Another recent study has claimed water in faults vaporizes during an earthquake, depositing gold. When an earthquake strikes, it moves along a fault. Water often lubricates faults, filling in fractures and jogs. About 6 miles (10 kilometres) below the surface, under incredible temperatures and pressures, the water carries high concentrations of carbon dioxide, silica, and gold. During an earthquake, the fault jog suddenly opens wider. The water inside the void instantly vaporizes, flashing to steam and forcing silica, which forms the mineral quartz, and gold out of the fluids and onto nearby surfaces.

Seawater

The world's oceans contain gold. Measured concentrations of gold in the Atlantic and Northeast Pacific are 50–150 fmol/L or 10–30 parts per 1,000,000,000,000,000 quadrillion (about 10–30 g/km^3). In general, gold concentrations for south Atlantic and central Pacific samples are the same (~50 fmol/L) but less certain. Mediterranean deep waters contain slightly higher concentrations of gold (100–150 fmol/L) attributed to wind-blown dust and/or rivers. At 10 parts per quadrillion the Earth's oceans would hold 15,000 tonnes of gold. These figures are three orders of magnitude less than reported in the literature prior to 1988, indicating contamination problems with the earlier data.

A number of people have claimed to be able to economically recover gold from sea water, but so far they have all been either mistaken or acted in an intentional deception. Prescott Jernegan ran a gold-from-seawater swindle in the United States in the 1890s. A British fraudster ran the same scam in England in the early 1900s. Fritz Haber (the German inventor of the Haber process) did research on the extraction of gold from sea water in an effort to help pay Germany's reparations following World War I. Based on the published values of 2 to 64 ppb of gold in seawater a commercially successful extraction seemed possible. After analysis of 4,000 water samples yielding an average of 0.004 ppb it became clear that the extraction would not be possible and he stopped the project. No commercially viable mechanism for performing gold extraction from sea water has yet been identified. Gold synthesis is not economically viable and is unlikely to become so in the foreseeable future.

Production

At the end of 2009, it was estimated that all the gold ever mined totaled 165,000 tonnes. This can be represented by a cube with an edge length of about 20.28 metres. At $1,600 per troy ounce, 165,000 metric tonnes of gold would have a value of $8.5 trillion.

World production for 2011 was at 2,700 tonnes, compared to 2,260 tonnes for 2008. Since the 1880s, South Africa has been the source for a large proportion of the world's gold supply, with about 50% of all gold ever produced having come from South Africa. Production in 1970 accounted for 79% of the world supply, producing about 1,480 tonnes.

In 2007 China (with 276 tonnes) overtook South Africa as the world's largest gold producer, the first time since 1905 that South Africa has not been the largest.

Mining

As of 2013, China was the world's leading gold-mining country, followed in order by Australia, the United States, Russia, and Peru. South Africa, which had dominated world gold production for most of the 20th Century, had declined to sixth place. Other major producers are the Ghana, Burkina Faso, Mali, Indonesia and Uzbekistan.

In South America, the controversial project Pascua Lama aims at exploitation of rich fields in the high mountains of Atacama Desert, at the border between Chile and Argentina.

Today about one-quarter of the world gold output is estimated to originate from artisanal or small scale mining. The city of Johannesburg located in South Africa was founded as a result of the Witwatersrand Gold Rush which resulted in the discovery of some of the largest gold deposits the world has ever seen. The gold fields are confined to the northern and north-western edges of the Witwatersrand basin, which is a 5–7 km thick layer of archean rocks located, in most places, deep under the Free State, Gauteng and surrounding provinces. These Witwatersrand rocks are exposed at the surface on the Witwatersrand, in and around Johannesburg, but also in isolated patches to the south-east and south-west of Johannesburg, as well as in an arc around the Vredefort Dome which lies close to the centre of the Witwatersrand basin. From these surface exposures the basin dips extensively, requiring some of the mining to occur at depths of nearly 4000 m, making them, especially the Savuka and TauTona mines to the south-west of Johannesburg, the deepest mines on earth. The gold is found only in six areas where archean rivers from the north and north-west formed extensive pebbly braided river deltas before draining into the "Witwatersrand sea" where the rest of the Witwatersrand sediments were deposited.

The Second Boer War of 1899–1901 between the British Empire and the Afrikaner Boers was at least partly over the rights of miners and possession of the gold wealth in South Africa.

Consumption

The consumption of gold produced in the world is about 50% in jewellery, 40% in investments, and 10% in industry. According to World Gold Council, China is the world's largest single consumer of gold in 2013 and toppled India for the first time with Chinese consumption increasing by 32 percent in a year, while that of India only rose by 13 percent and world consumption rose by 21 percent. Unlike India where gold is used for mainly for jewellery, China uses gold for manufacturing and retail.

Resource Depletion

Resource depletion is the consumption of a resource faster than it can be replenished. Resources are commonly divided between renewable resources and non-renewable resources. Use of either of these forms of resources beyond their rate of replacement is considered to be resource depletion. Resource depletion is most commonly used in reference to farming, fishing, mining, water, and fossil fuels.

Causes

- Overconsumption; excessive or unnecessary use of resources
- Overpopulation
- Slash and burn agricultural practices, currently occurring in many developing countries
- Technological and industrial development
- Erosion
- Habitat degradation leads to the loss of Biodiversity (i.e. species and ecosystems with its ecosystem services).
- Irrigation
- Mining for oil and minerals
- Aquifer depletion
- Pollution or contamination of resources

Minerals

Minerals are needed to provide food, clothing, and housing. A USGS study found a significant long-term trend over the 20th century for nonrenewable resources such as minerals to supply a greater proportion of the raw material inputs to the non-fuel, non-food sector of the economy; an example is the greater consumption of crushed stone, sand, and gravel used in construction.

Large-scale exploitation of minerals began in the Industrial Revolution around 1760 in England and has grown rapidly ever since. Most of the world's mineral ores are still being extracted from mines over fifty years old. Miners cope by digging deeper, accepting lower grades of ore, and using technology to extract the minerals. Virtually all basic industrial metals (copper, iron, bauxite, etc.), as well as rare earth minerals, are facing output limitations. Minerals projected to enter production decline during the next 20 years:

- Gas (2023)
- Copper (2024)
- Zinc

Minerals projected to enter production decline during the present century:

- Coal? (2030-2060) 2060
- Iron (2068)
- Aluminium (2057)

Oil

Peak oil is the period when the maximum rate of global petroleum extraction is reached, after which the rate of production enters terminal decline. It relates to a long-term decline in the available supply of petroleum. This, combined with increasing demand, will significantly increase the worldwide prices of petroleum derived products. Most significant will be the availability and price of liquid fuel for transportation. The US Department of Energy in the Hirsch report indicates that "The problems associated with world oil production peaking will not be temporary, and past "energy crisis" experience will provide relatively little guidance."

Deforestation

Deforestation is the clearing of natural forests by logging or burning of trees and plants in a forested area. As a result of deforestation, presently about one half of the forests that once covered the Earth have been destroyed. It occurs for many different reasons, and it has several negative implications on the atmosphere and the quality of the land in and surrounding the forest.

Causes

One of the main causes of deforestation is clearing forests for agricultural reasons. As the population of developing areas, especially near rainforests, increases, the need for land for farming becomes more and more important. For most people, a forest has no value when its resources aren't being used, so the incentives to deforest these areas outweigh the incentives to preserve the forests. For this reason, the economic value of the forests is very important for developing worlds.

Environmental Impact

Because deforestation is so extensive, it has made several significant impacts on the environment, including carbon dioxide in the atmosphere, changing the water cycle, an increase in soil erosion, and a decrease in biodiversity. Deforestation is often cited as a cause of global warming. Because trees and plants remove carbon dioxide and emit oxygen into the atmosphere, the reduction of forests contribute to about 12% of anthropogenic carbon dioxide emissions.

One of the most pressing issues that deforestation creates is soil erosion. The removal of trees causes higher rates of erosion, increasing risks of landslides, which is a direct threat to many people living close to deforested areas.

As forests get destroyed, so does the habitat for millions of animals. It is estimated that 80% of the world's known biodiversity lives in the rainforests, and the destruction of these rainforests is accelerating extinction at an alarming rate.

Controlling Deforestation

The United Nations and the World Bank created programs such as Reducing Emissions from Deforestation and Forest Degradation (REDD), which works especially with developing countries to use subsidies or other incentives to encourage citizens to use the forest in a more sustainable way. In addition to making sure that emissions from deforestation are kept to a minimum, an effort to educate people on sustainability and helping them to focus on the long-term risks is key to the success of these programs. Reforestation is also being encouraged in many countries in an attempt to repair the damage that deforestation has done.

Wetlands

Causes

- Overconsumption; excessive or unnecessary use of resources
- Overpopulation
- Slash and burn agricultural practices, currently occurring in many developing countries
- Technological and industrial development
- Erosion
- Habitat degradation leads to the loss of Biodiversity (i.e. species and ecosystems with its ecosystem services).
- Irrigation
- Mining for oil and minerals
- Aquifer depletion
- Pollution or contamination of resources

Minerals

Minerals are needed to provide food, clothing, and housing. A USGS study found a significant long-term trend over the 20th century for nonrenewable resources such as minerals to supply a greater proportion of the raw material inputs to the non-fuel, non-food sector of the economy; an example is the greater consumption of crushed stone, sand, and gravel used in construction.

Large-scale exploitation of minerals began in the Industrial Revolution around 1760 in England and has grown rapidly ever since. Most of the world's mineral ores are still being extracted from mines over fifty years old. Miners cope by digging deeper, accepting lower grades of ore, and using technology to extract the minerals. Virtually all basic industrial metals (copper, iron, bauxite, etc.), as well as rare earth minerals, are facing output limitations.

Minerals projected to enter production decline during the next 20 years:

- Gas (2023)
- Copper (2024)
- Zinc

Minerals projected to enter production decline during the present century:

- Coal? (2030-2060) 2060
- Iron (2068)
- Aluminium (2057)

Oil

Peak oil is the period when the maximum rate of global petroleum extraction is reached, after which the rate of production enters terminal decline. It relates to a long-term decline in the available supply of petroleum. This, combined with increasing demand, will significantly increase the worldwide prices of petroleum derived products. Most significant will be the availability and price of liquid fuel for transportation.

The US Department of Energy in the Hirsch report indicates that "The problems associated with world oil production peaking will not be temporary, and past "energy crisis" experience will provide relatively little guidance."

Deforestation

Deforestation is the clearing of natural forests by logging or burning of trees and plants in a forested area. As a result of deforestation, presently about one half of the forests that once covered the Earth have been destroyed. It occurs for many different reasons, and it has several negative implications on the atmosphere and the quality of the land in and surrounding the forest.

Causes

One of the main causes of deforestation is clearing forests for agricultural reasons. As the population of developing areas, especially near rainforests, increases, the need for land for farming becomes more and more important. For most people, a forest has no value when its resources aren't being used, so the incentives to deforest these areas outweigh the incentives to preserve the forests. For this reason, the economic value of the forests is very important for developing worlds.

Environmental Impact

Because deforestation is so extensive, it has made several significant impacts on the environment, including carbon dioxide in the atmosphere, changing the water cycle, an increase in soil erosion, and a decrease in biodiversity. Deforestation is often cited as a cause of global warming. Because trees and plants remove carbon dioxide and emit oxygen into the atmosphere, the reduction of forests contribute to about 12% of anthropogenic carbon dioxide emissions. One of the most pressing issues that deforestation creates is soil erosion. The removal of trees causes higher rates of erosion, increasing risks of landslides, which is a direct threat to many people living close to deforested areas. As forests get destroyed, so does the habitat for millions of animals. It is estimated that 80% of the world's known biodiversity lives in the rainforests, and the destruction of these rainforests is accelerating extinction at an alarming rate.

Controlling Deforestation

The United Nations and the World Bank created programs such as Reducing Emissions from Deforestation and Forest Degradation (REDD), which works especially with developing countries to use subsidies or other incentives to encourage citizens to use the forest in a more sustainable way. In addition to making sure that emissions from deforestation are kept to a minimum, an effort to educate people on sustainability and helping them to focus on the long-term risks is key to the success of these programs. Reforestation is also being encouraged in many countries in an attempt to repair the damage that deforestation has done.

Bibliography

Aguilera, Jose Miguel and David W. Stanley: *Microstructural Principles of Food Processing and Engineering*: Springer, US., 2004.

Ben-Naim, A. and Ben-Naim, R., P.H.: *Alice's Adventures in Water-land*: World Scientific Publishing, UK, 2011.

Berry, John: *Tangible Strategies for Intangible Assets*, McGraw-Hill, United States, 2004.

Buol, S. W.; Hole, F. D.; McCracken, R. J.: *Soil Genesis and Classification*, Iowa State University Press, Ames, Iowa, 1973.

Campbell, Bernard Grant: *Human Evolution: An Introduction to Man's Adaptations*, Aldine Transaction, NJ, 1998.

Cantril, T. C.: Coal Mining, Cambridge University Press, Cambridge, England, 1914.

Chesworth, Ward: *Encyclopedia of soil science*, Springer, Dordrecht, Netherlands, 2008.

Dalrymple, G.B.: *The Age of the Earth*, Stanford University Press, California, 1991.

Davidson, Alan: *The Oxford Companion to Food*, Oxford University Press, UK, 2006.

Forbes, RJ: *Studies in Ancient Technology*, Brill Academic Publishers, Boston, 1966.

Foth, Henry D.: *Fundamentals of soil science*, Wiley, New York, 1984.

Gleick, P.H.: *Water in Crisis: A Guide to the World's Freshwater Resources*, Oxford University Press, 1993.

Humphery, Kim: *Shelf Life: Supermarkets and the Changing Cultures of Consumption*, Cambridge University Press, 1998.

Jango-Cohen, Judith: *The History of Food*, Twenty-First Century Books, United States, 2005.

Kotz, J. C., Treichel, P., & Weaver, G. C.: *Chemistry & Chemical Reactivity*, Thomson Brooks/Cole, US., 2005.

Leake, Simon; Haege, Elke: *Soils for Landscape Development*, CSIRO Publishing, Melbourne, Australia, 2014.

Lomborg, Björn: *The Skeptical Environmentalist*, Cambridge University Press, London, 2001.

Mankiw, N.G.: *Principles of Economics*, South-Western College Publishing, Boston, MA., 2008.

McCarthy, David F.: *Essentials of Soil Mechanics and Foundations: Basic Geotechnics,* Reston Publishing, Reston, Virginia, 1982.

Miller, G.T., and S. Spoolman: *Living in the Environment: Principles, Connections, and Solutions,* Brooks-Cole, Belmont, CA, 2011.

Nyle C. Brady & Ray R. Weil: *Elements of the Nature and Properties of Soils,* Prentice Hall, United States, 2009.

Nestle, Marion: *Food Politics: How the Food Industry Influences Nutrition and Health,* University Presses of California, 2007.

Outwater, Alice: *Water: A Natural History,* Basic Books, New York, NY, 1996.

Parekh, Sarad R.: *The Gmo Handbook: Genetically Modified Animals, Microbes, and Plants in Biotechnology*, Humana Press, New York, 2004.

Paul, E.A.: *Soil organic matter in temperate agroecosystems : long-term experiments in North America,* CRC Press, Boca Raton, 1997.

Retallack, G. J.: *Soils of the past : an introduction to paleopedology*, Unwin Hyman, Boston, 1990.

Retallack, G.J.: *Soils of the Past: An Introduction to Paleopedology*, John Wiley & Sons, 2008.

Ricklefs, R.E.: *The Economy of Nature,* WH Freeman, New York, NY, 2005.

Rottenberg, Dan: *In the Kingdom of Coal; An American Family and the Rock That Changed the World,* Routledge, US, 2003.

Sanchez, Pedro A.: *Properties and management of soils in the tropics*, Wiley, New York, 1976.

Schor, Juliet; Taylor, Betsy: *Sustainable Planet: Roadmaps for the Twenty-First Century*, Beacon Press, 2003.

Van den Bossche, Peter: *The Law and Policy of the bosanac Trade Organization: Text, Cases and Materials*, Cambridge University Press, UK, 2005.

Voroney, R. P.: Soil Microbiology, Ecology and Biochemistry, Cambridge University Press, UK, 2006.

Walter Licht, Thomas Dublin: *The Face of Decline: The Pennsylvania Anthracite Region in the Twentieth Century*, Cornell University Press, United States, 2005.

Wrigley, EA.: *Continuity, Chance and Change: The Character of the Industrial Revolution in England*, Cambridge University Press, 1990.

Index

M

N

O

P

R

S

T

V

W

❑❑❑